Routledge Revivals

U.S. Interests and Global Natural Resources

Originally published in 1983, *U.S. Interests and Global Natural Resources* explores the links between foreign policy and the global distribution of natural resources paying particular attention to the U.S. This collection of essays delves into the importance of factors such as differing economic development and political hostility could have on the provision of resources into the U.S and advises that nations identify their countries needs and establish policies to safeguard them. This title will be of interest to students of Environmental Studies and Policy makers.

U.S. Interests and Global Natural Resources

Energy, Minerals, Food

Edited By
Emery N. Castle and Kent A. Price

RFF PRESS
RESOURCES FOR THE FUTURE

First published in 1983
by Resources for the Future, Inc.

This edition first published in 2016 by Routledge
2 Park Square, Milton Park, Abingdon, Oxon, OX14 4RN
and by Routledge
711 Third Avenue, New York, NY 10017

Routledge is an imprint of the Taylor & Francis Group, an informa business

Publisher's Note

The publisher has gone to great lengths to ensure the quality of this reprint but points out that some imperfections in the original copies may be apparent.

Disclaimer

The publisher has made every effort to trace copyright holders and welcomes correspondence from those they have been unable to contact.

A Library of Congress record exists under LC control number: 83042905

ISBN 13: 978-1-138-94697-2 (hbk)
ISBN 13: 978-1-315-67034-8 (ebk)

U.S. INTERESTS AND GLOBAL NATURAL RESOURCES

energy, minerals, food

edited by
EMERY N. CASTLE
KENT A. PRICE

Published by Resources for the Future, Inc.
Washington, D.C.

Distributed by The Johns Hopkins University Press
Baltimore and London

Published by Resources for the Future, Inc., 1755 Massachusetts Avenue, N.W., Washington, D.C. 20036
Distributed by The Johns Hopkins University Press, Baltimore, Maryland 21218
The Johns Hopkins Press Ltd., London.

Library of Congress Cataloging in Publication Data
Main entry under title:

U.S. interests and global natural resources.

Includes index.
1. Natural resources—Government policy—United States. 2. Strategic materials—United States. 3. United States—Foreign relations. 4. United States—Foreign economic relations. 5. United States—National security.
I. Castle, Emery N. II. Price, Kent A. III. Title: US interests and global natural resources. IV. Title: United States interests and global natural resources.
HC103.7.U18 1983 333.7'0973 83-42905

ISBN 0-8018-3099-0
ISBN 0-8018-3106-7 (pbk.)

Resources for the Future is a nonprofit organization for research and education in the development, conservation, and use of natural resources, including the quality of the environment. It was established in 1952 with the cooperation of the Ford Foundation. Grants for research are accepted from government and private sources only on the condition that RFF shall be solely responsible for the conduct of the research and free to make its results available to the public. Most of the work of Resources for the Future is carried out by its resident staff; part is supported by grants to universities and other nonprofit organizations. Unless otherwise stated, interpretations and conclusions in RFF publications are those of the authors; the organization takes responsibility for the selection of significant subjects for study, the competence of the researchers, and their freedom of inquiry.

This book was edited by Jo Hinkel and designed by Elsa B. Williams. The index was prepared by Florence Robinson. The figures were drawn by Arts and Words.

Contents

Tables and Figures

Tables

Figures

Preface

If there ever was a golden age when the United States could indulge a dream of economic self-sufficiency, that time has long since passed. For example:

- U.S. oil imports have declined dramatically from their 1977 peak, but still flow at a rate of more than 4 million barrels every day. The price tag exceeds $40 billion annually and is incalculable in terms of national security.
- The United States relies heavily on foreign sources—some of them of questionable stability—for a long list of nonfuel minerals. At least for the short run, a few of these minerals are critical ingredients of a high-technology economy.
- Fully one-third of the product of American agriculture is exported to other countries. The United States accounts for nearly 60 percent of the world's total grain trade and exports 10 percent of the world's total consumption of wheat.

International trade in natural resources consists of interwoven threads of demand and supply: that one country's need to buy is another's opportunity to sell makes for a world of mutual dependence and relative stability. But nature did not distribute all resources equally around the globe, and the fact that different societies are at differing stages of economic development further complicates the random distribution of petroleum, or chromium, or temperate climate. Geographic or economic dominance of a market can lead to cartels, market strength coupled with political hostility makes embargoes possible, natural disasters can shut off resource supplies, and wars, revolutions, and terrorism often place the links of trade high on the list of targets.

In such a world, a nation is well advised to identify its vital interests and to develop and establish policies that safeguard those interests. In the fall of 1982, Resources for the Future convened a Forum to define U.S. national interests and examine policies concerning three groups of natural resources that are conspicuous by their importance—energy, minerals, and food. This volume is a result of that gathering.

Policy Gaps

At least since the early 1970s, the supply, demand, and national security aspects of oil have been central influences on U.S. foreign policy. Similarly, U.S. dependence on foreign sources for so-called strategic minerals and enormous American agricultural exports have undeniable impact on the nation's conduct of its commerce and its foreign affairs.

Thus, U.S. natural resource policy should not be formulated in isolation from world events. The United States has become increasingly interdependent with other countries through trade in natural resource commodities. This does not mean that America is a resource-poor country; the facts provide no support for any such conclusion. Rather, it flows from the twin realities that natural resources are neither uniformly distributed nor equally accessible in all locations. Consequently, grain embargoes, the availability of minerals, oil imports, and transnational pipelines increasingly have become the substance of foreign policy. Yet, natural resource policy too frequently is formulated on a resource-by-resource basis, with the key factors being domestic interests. For example, the principal U.S. resource agency is called the Department of the *Interior*. Certainly too much could be made of this, but my point is that we need to look outward as well as inward when natural resource policy is formulated. Conversely, U.S. foreign policy and deliberations about international relations typically consider natural resource issues only on an *ad hoc* or episodic basis. Accordingly, our objective at the Forum and in this book is to contribute to the integration of these two major strands of national concern.

Plan of the Book

The opening chapter, by my co-editor and me, introduces the subject and explores several of the key ideas presented by the other authors. The succeeding chapters follow the order of the Forum presentations.

In chapter two, Milton Russell, assisted by Mary Beth Zimmerman, addresses the topic of energy by focusing on the key energy commodity—oil. Using the dramatic oil revolution of the 1970s as his primary case study, Russell lays out the realities of global supply and demand, assesses the implications for U.S. economic policy, and discusses both national and international aspects of energy security policy.

Chapter three explores the facts and fears of U.S. dependence on foreign sources—some of them unstable—for several key nonfuel minerals. John E. Tilton and Hans H. Landsberg provide a useful framework and perspective for assessing the dilemmas posed by imports for the United States and the extent to which U.S. policy can respond effectively.

Dale E. Hathaway, in chapter four, outlines the startling and probably unique developments of the 1970s regarding U.S. agriculture and its links to foreign markets. He argues that U.S. agriculture has been thoroughly internationalized. That is, domestic economic pressures relating to food are exerting increasing influence on the way that the United States approaches international affairs, and domestic markets and policies increasingly are driven by foreign markets and international events.

Whereas chapters two, three, and four are primarily resource commodity based, Carl Kaysen in chapter five focuses directly on U.S. foreign policy. To what extent, he asks, is foreign policy made—or how should it be made—in light of the worldwide distribution and availability of natural resources and their services? Readers may find Kaysen's historical perspective particularly helpful in assessing today's problems and opportunities.

McGeorge Bundy, the final speaker at the Forum, offers instructive comments on the earlier presentations and contributes the insights of a key participant in the making of U.S. foreign policy, and in evaluating it as well. His eloquent remarks are included here as an epilogue.

The Developing World

When we were planning the Forum program we were acutely aware that, despite our focus on U.S. national interests, the United States could not be considered an island in a global sea; indeed, a major purpose and theme of the Forum was to demonstrate the ineradicable links among all the nations of the world. In this regard we were especially conscious of the growing and significant role played individually and collectively by the developing nations. Depending on the resources involved, some are exporters and some are importers, but all are affected in major ways by shifts in the availability and demand for resources. It is critical that the developing country point of view be kept in mind as we consider natural resources and international relations.

Accordingly, we asked Raj Krishna of the University of Delhi and Stanford University to be one of the major Forum speakers. Unfortunately, medical problems forced Dr. Krishna to withdraw, and both the Forum and this volume are the poorer for the lack of the perspective he would have brought to the subject. However, in a panel discussion following the major presentations, John W. Sewell, president of the Overseas Development Council, spoke extemporaneously about the developing societies. His comments then were especially helpful in fleshing out a complete picture of the topics under consideration, so much so that an edited portion of his remarks is presented here.

> *The growing importance of the Third World.* In 1959 there were 92 independent countries. Now there are well over 160, and all of the new ones are self-described members of the developing world. Numbers alone, of course, tell only a partial story. Most developing country spokesmen feel very strongly that their importance—their economic and political importance—to the well-being of the industrial world has grown significantly over the last two decades. Indeed, during that time the economic growth rate of many countries in the

Third World has far exceeded those of countries in the industrial world. Nor are these only the newly rich members of the Organization of Petroleum Exporting Countries. Even leaving aside OPEC, developing countries now buy more from the United States than does the European Economic Community, Eastern Europe, the Soviet Union, and China combined. They also have been the fastest-growing export market for the United States, and their role in the international financial system—whether for stability or instability—needs no emphasis to anyone who has been following events in Mexico or Brazil or elsewhere among the major debtors in the Third World.

In the political realm, there is a growing sense of grievance in the Third World that the economic and political systems created after World War II, the Bretton Woods system and its ancillary organizations, give them neither an adequate share of benefits nor of power. They do not insist on a dogmatic adherence to a one-nation, one-vote requirement, but they do not believe the present systems reflect their much more important participation in the international economic and political arenas either in distribution of power or of benefits.

Resource and trade issues. Despite the growing importance of the Third World in producing manufactured goods, well over half the non-OPEC developing countries still depend for over 50 percent of their total exports on two or three agricultural or non-oil raw materials. The issues of commodity price stabilization and improvements in the terms of trade therefore still are very important to the Third World.

We have seen an historic shift: major areas of the developing world were net food exporters before World War II and now are very large food importers. The real issue for most of the developing countries is how they can improve their own food production; and that they do so is also very much in the U.S. interest. A greater degree of food self-reliance is important both for political and economic reasons, as well as for human reasons in addressing the problems of poverty. The question in most of the Third World countries is why the industrial world is not being more helpful in this regard, either on price stabilization measures or on assistance to agricultural production.

On the issue of oil, it is true that heavy costs were extracted from the oil-importing developing countries since 1974, but they have put much less political pressure on OPEC than on the industrial countries, for two reasons. The first is OPEC aid: until recently, OPEC aid as a percentage of its members' gross national products was well above Organisation for Economic Co-operation and Development aid as a percentage of GNP. And concessional oil sales have been quite remarkable. The second reason is more psychological than economic: there still is a very strong sense of Third World solidarity.

In contrast, there is a seeming lack of interest in the industrial countries in assistance for energy production in the Third World, which generates resentment. Also, political pressure is exercised because Third World leaders frankly recognize that the major gains are to be made in discussions with the industrial world: the OECD countries still dominate the world's economy.

U.S. policy. What do these issues mean for American foreign policy? When one looks at U.S. policy toward the developing countries, one sees a marked disparity between the real economic, political, and social interests of the United States and the low priorities that public and private decision makers give to these countries. I would argue that this is going to incur increasing costs over the long term for the United States, though it is easy to understand why the Third World is relatively slighted. U.S. foreign policy is dominated by four axioms, one of which is an argument about the intentions of the Soviet Union, which generally are considered to be malign. The second is a concern about instability in the Third World, an instability considered to be dangerous for the United States, either in itself or because it opens opportunities for the Soviet Union. The third concerns the utility of military power and U.S. ability to project that power into the developing world. And the last is the future of the Atlantic alliance and the primacy that it is accorded in American foreign policy.

None of these axioms is unimportant. But none of them will be adequate, it seems to me, in the next two decades when the developing countries are going to be much more important to the United States over a whole range of national interests. I certainly would not propose that Americans ignore the very real human problems that plague the Third World, but it perhaps is important to stress that traditional American sympathy for the underdog need not necessarily play a role in U.S. relations with the developing societies. Strictly in terms of *Realpolitik,* the United States now has a direct national interest in what happens in terms of growth and progress in the developing countries. Whatever direction these countries take in the future is going to have a great impact on well-being in the United States and consequently on U.S. foreign policy.

The nations of the developing world are considered both explicitly and implicitly throughout the book, but I thought it appropriate to emphasize their importance at the outset. John Sewell's points are well taken.

Acknowledgments

What follows in a sense are the proceedings of the RFF Forum held on October 21, 1982, in Washington, D.C. But it is more than a proceedings volume, for each of the chapters is considerably more comprehensive and detailed than is possible in oral presentation, and in nearly every case the authors revised their work and brought it up to date. Thus, I am grateful to the authors not only for their spirited presentations to the Forum, but also for their extra effort toward making this a book that stands on its own merits.

In this regard, I also want to mention the contributions of my co-editor, Kent A. Price. He integrated varying styles, assured consistency, edited substance, suggested revisions, and in general was instrumental in transforming a collection of papers into a cohesive volume.

Elizabeth E. Davis, research assistant in RFF's Center for Energy Policy Research, carefully read and verified chapter two. I join Milton Russell in expressing to her our appreciation for minimizing any errors of fact or interpretation. Thanks also are due D. J. Curran, whose last-minute word processing kept the manuscript on schedule at a critical point.

The logistics of producing even a relatively slim volume are enormously burdensome. Much of this was borne with meticulous skill and considerable good humor by Patricia M. Parker, who spent countless hours at her word processing terminal to assure a product of the highest possible quality. In this book, as in many others, RFF editor Jo Hinkel was responsible for dozens of details and loose ends that, left to less competent hands, might be half-checked and untied. The book was designed by Elsa B. Williams, who also—as RFF's production editor—oversaw the typesetting, printing, and binding operations with her usual deft touch. RFF is fortunate indeed to have such a trio on the publications front line.

Washington, D.C.
August 1983

Emery N. Castle
President and Senior Fellow
Resources for the Future

chapter one

The International Dimension

Emery N. Castle
Kent A. Price

The practice of international statecraft always has been a precarious occupation: relations among nations can be thrown off course by a myriad of factors. Moreover, the forces contributing to instability have increased substantially since World War II. No secretary of state since has had the luxury of remaining in Washington and mapping grand strategy with his president. While a few may have entered office with some such romantic notion, all have been forced to spend much of their time traveling to sensitive parts of the globe to maintain existing alliances or to establish new ones. And much the same is true of foreign ministers of other world powers.

In a way, of course, this is reassuring. Not all grand designs for the world have the preservation of freedom and peace as their primary objectives. And with power, wealth, and influence dispersed as they are, no monolithic power can impose its will on any significant part of the globe without allies. Allies and alliances by definition are creatures of some common objective or interest. But those common interests and objectives often disappear under the pressure of autonomous forces that affect allies and opponents in different ways. For example, the United States and many Western European countries, staunch allies since World War II, find their interests diverge when they try to increase agricultural exports simultaneously. The United States and the Soviet Union may be long-time antagonists but discover that it is in their common interest for one to export and the other to import large quantities of food and feed grains, or perhaps even technology.

During the past decade new destabilizing forces have harassed international relations. Early in the 1970s the petroleum-exporting countries recognized the international market power they could exert by acting in concert, with enormous impacts on international relations and the distribution of wealth. The United States was forced to rethink and redefine its relations, not just with the oil exporters, but also with the Soviet Union and other centrally planned economies and with the great majority of developing countries that are oil importers. Further, the oil price revolution influenced U.S. relations with a long-time friend, Israel.

Changes in energy prices and availability have been the most dramatic new destabilizing forces, but they by no means are the only ways that natural resources and the natural environment influence international relations. Changing consumption, production, and trading patterns in food affect national diets and fortunes. Environmental problems such as acid rain and carbon dioxide accumulation in the atmosphere cut across national boundaries with bilateral and even global implications. Rhetoric about alleged Soviet-sponsored "resource wars" and U.S. dependence on foreign sources of minerals have been used to justify the exploration and development of land in the United States that may be better devoted to wilderness and other purposes. In short, resource and environmental issues have emerged as major unsettling forces in international relations, and we see little prospect that this situation will change anytime soon. Indeed, quite the contrary is likely. Especially as far as the United States is concerned, it would be surprising if U.S. interests as they affect and are affected by global natural resources do not assume a central position on the national agenda in the decades to come.

How well does U.S. foreign policy take into account the global distribution of natural resources? Conversely, how well does U.S. domestic policy reflect the reality that we share one globe and that the interests of American producers of natural resource services must be balanced against the objectives of U.S. foreign policy? We believe that considerable improvement is possible. The availability, production, and distribution of natural resources and their services often are neglected when foreign policy is discussed and made. On the other side of the coin, natural resource policies typically are perceived as exclusively domestic; they are dominated by the interests of domestic producers, and U.S. international objectives are all but ignored except in times of emergency. When an emergency does arise, overreactions such as embargoes frequently cause enormous adjustments and dislocations in both importing and exporting countries.

Resource Differences

The chapters to follow explore the production, distribution, use, and control of the natural resources of energy, minerals, and food, and examine how this knowledge could and should affect U.S. foreign policy. Perhaps the overriding conclusion to emerge is that the realities of these global resources affect the United States in very different ways. Certain common principles govern discovery, development, use, and trade in natural resources and natural resource services, but resource-specific differences are so sharp that a comprehensive natural resource foreign policy is impossible. Thus, two fundamental sets of propositions must be addressed if the necessary knowledge about natural resources is to be incorporated into the U.S. approach to the rest of the world. First, basic knowledge and principles about the natural resources themselves, including their distribution around the globe, must be made explicit. Second, the actual and potential position of the United States relative to these resources needs to be specified.

The United States is the world's second largest producer of oil, after the Soviet Union, but the prodigious American appetite for oil makes the country heavily dependent on foreign sources of petroleum, several of which are concentrated in the Middle East. Imported oil is profoundly important to the U.S. economy in the aggregate and as it affects many individual production and consumption activities. Liquid fuels, which oil dominates, are uniquely useful, especially in transportation, and the U.S. national annual bill for them—in excess of $40 billion for foreign oil alone and twice as much only three or four years ago—reflects that value. Substitutes for oil are few and not now available on anything like the scale that would be needed were oil to disappear.

The situation is very different in the case of nonfuel minerals. Compared to the single petroleum basic resource—crude oil—there are many kinds of minerals, they are used in numerous ways in the economy, and U.S. dependence on foreign sources varies greatly. Moreover, the values involved are in totally different ballparks: the annual U.S. cost for imported cobalt, chromium, manganese, or platinum varies from $250 million to $500 million, less than what the country has recently been spending in two days for imported petroleum. Further, the development of substitutes ranges from certain to possible if imports were cut off from countries likely to become unfriendly to the United States or otherwise lost as suppliers.

Food presents still another picture. The United States exports about one-third of the food it produces and would like to export even more. Perhaps erroneously reasoning from the example of the oil exporters, some believe that food exports constitute significant U.S. leverage over the centrally planned economies which are major importers of American agricultural products. But less than happy recent experience with U.S. embargoes makes one skeptical about such a proposition. Indeed, the United States neither produces itself nor otherwise controls enough existing sources of grain, let alone potential production, to make an effective food cartel possible.

Milton Russell calls our attention in chapter two to an obvious but often overlooked point concerning differing perceptions of the same phenomenon. The oil price increases of the 1970s were viewed with anger by the oil-importing countries as conveying enormous—even vaguely criminal— windfall gains to the exporting countries. Of course, massive gains did occur, but the exporting countries see them as if through the other end of a telescope: aware of the finiteness of their oil resources, they view exports as the disappearance of an irreplaceable birthright for which they had been shortchanged for years. The potential for differing perceptions is large. For example, while an oil importer may worry about the effect that higher oil prices have on inflation and economic stagnation, an exporting country may be anxious about the rising expectations of its people. While the United States may see grain exports to the centrally planned economies as conferring leverage and power, the view may be quite different in the importing countries. It may be more economic for them to import than to produce, but their leaders can afford to treat an interruption of food imports as an annoyance rather than as a catastrophe. They can reduce consumption or turn to another source of supply rather than accept a major accommodation or shift in foreign policy objectives. Countries understandably may be concerned about depending on imported oil or food, but exporting countries may have comparable apprehensions about depending on the same transaction for foreign exchange. Differing perceptions notwithstanding, however, one fact is inescapable: trade in natural resources or natural resource services does create a degree of interdependence between exporter and importer. Exchange freely arrived at over time will benefit both parties.

Cartel Fears Unfounded

Much of the apprehension about imports stems from the fear that market power can be concentrated sufficiently to limit or interrupt the flow or significantly raise the price of a critical resource. No doubt the success of the Organization of Petroleum Exporting Countries during the 1970s has contributed greatly to this apprehension. But major exporters also may wonder if they could extract a greater reward for their products or perhaps use the threat of a supply interruption to achieve a foreign policy objective. John Tilton and Hans Landsberg in chapter three provide a framework for evaluating the possible political power of those providing for the flow of natural resources or natural resource services; they outline four conditions necessary for a successful cartel:

- The number of major producing and exporting countries must be limited.
- Sales losses resulting from price rises must be small, or the advantage gained by raising prices will be more than lost by reduced volume.
- Price rises by the cartel must not generate much additional production by those outside it.
- The bonds among cartel members must be strong enough to prevent desertions or cheating.

Tilton and Landsberg offer these four points in the context of nonfuel minerals, but they are equally helpful in thinking about other resources and resource services including energy and food. Experience reveals another truth common to all resource cartels, namely, that the length of time involved makes a great deal of difference. For example, in the short run only established producers may be effective market participants, but eventually potential suppliers are able to come into the market. Natural resources are unevenly distributed around the world, but no small area or region has a monopoly on any resource or its close substitutes.

Thus we see why natural resource issues became so important in international relations during the 1970s. It is clear that because petroleum-producing capacity was sufficiently concentrated, and that the industrial world had so few close substitutes for oil, that a cartel could be effective—for a while. However, with the passage of time OPEC's power has diminished. If oil embargoes are threatened or imposed in the future, their effectiveness will depend in large part on the extent to which oil-importing countries have prepared themselves for such a contingency, on the future concentration in petroleum production, and on the growth of alternative sources of energy.

The same logic helps to explain why a crisis in the nonfuel minerals has not developed despite periodic gloomy predictions. Further, it suggests that fears of "resource wars" are likely to be unfounded.

Some American mistakes in international food policy also can be traced to a failure to appreciate the significance of the four points developed above. As noted, the rapid growth in U.S. food exports, detailed by Dale Hathaway in chapter four, has been taken by some as evidence of significant American market power in international grain markets. True, the Soviet Union and the other centrally planned economies have made major mistakes in agriculture and do need to import grains if the standard of living of their people is to continue to rise. Further, it is to their advantage to buy large amounts from the United States because it is a major efficient producer. But this does not necessarily translate into market or political power for the United States. The centrally planned economies can make two significant adjustments to an interruption of grain imports from the United States, and they have done so.

First, they can cultivate alternative sources of supplies. While American influence in the world is substantial, there is no prospect that the United States can control the exporting policies of all existing and potential grain producers. The United States is very efficient in the production of both food and feed grains, but many other places in the world can increase the output of these commodities if an effective demand exists. Second, the importing countries have great capacity to adjust their consumption of grains. A large part of the grain imports into the centrally planned economies is used for livestock feed. The trend toward more animal products in the diet is associated with a rising standard of living, and the leaders of such countries would like to be able to continue the trend. But their imports in this sense are a luxury: they are not necessary to prevent starvation, nor are they usually necessary to maintain social stability. In a society inured to hardship and responsive to authoritarian direction, it usually is possible to tighten the national belt another notch. This is especially true if blame can be cast on a country's ideological opponents.

Lessons for the Future

Thus, careful study of recent experience reveals that the capacity to export nonfuel minerals or food does not carry with it a great deal of international power or influence. Even in the unique case of petroleum, the power conferred by exports is likely to be transitory and unstable. Relations among nations should reflect natural resource realities and, in particular, these realities should become an integral part of U.S. foreign policy. But any advantage enjoyed by a nation or a group of nations because they are major exporters at a point in time almost certainly is illusory and temporary and unlikely to be effective in furthering national interests if used as a blunt instrument. To the contrary: aggressive exploitation of advantage will generate forces that erode and typically, in time, eliminate that advantage.

Natural resource and environmental issues will be an integral part of international relations for as far ahead as planning is relevant. Natural resources are not distributed evenly over the face of the globe and access to these resources and their services will occupy the attention of many nations. The international management of the open-access resources, such as the atmosphere and the oceans, provides the opportunity—indeed, the necessity—for international cooperation.

Yet the U.S. government is not well structured to permit natural resources and environmental quality to be considered in foreign policy deliberations, nor for foreign policy objectives to be reflected in natural resource or environmental policy. The Departments of Agriculture, Interior, Energy, and the Environmental Protection Agency understandably must be concerned with partial views, domestic affairs, and domestic producer and consumer interests. The Department of State and the National Security Council apparatus have difficulty incorporating subject matter knowledge and domestic political considerations in foreign policy. The institutional base for the formulation of foreign policy needs to reflect both the enduring nature and the importance of natural resource and environmental matters.

The role of domestic producers in influencing foreign policy is common to all resources. It is appropriate that a country's foreign policy protect the welfare of its citizens in their dealings with other nations. Yet, the strategic interests of a nation are not necessarily synonymous with improving the economic welfare of domestic producers. But, except in times of emergency, U.S. trade policy objectives seem to be preoccupied with the latter. It is possible to do better.

International trade in natural resource commodities and services creates greater international interdependence, which is a long-run influence for peace. Although the market power of one country or a bloc of countries usually cannot be translated into permanent political power, international trade does create interests in both exporting and importing countries that benefit from peace and stable trade relations. Thus, as Carl Kaysen concludes in chapter five, transnational pipelines, U.S. grain exports to the Soviet Union, and U.S. imports of minerals from a variety of sources are positive, cooperative developments that augur well for the future. An increasingly complex pattern of trade fosters mutual dependence and increases the costs of hostility.

A competitive trading pattern in natural resource commodities and services requires a degree of stability in international relations. As Kaysen notes, the United States does not have a long tradition as a major exporter. If America aspires to maintain and enhance its role as an exporter, it needs to establish a reputation as a reliable supplier. This suggests that destabilizing devices, such as embargoes, be used sparingly and judiciously and certainly not as a means of achieving immediate objectives at the expense of strategic interests.

Expect the Unexpected

Finally, as McGeorge Bundy emphasizes in the epilogue, it is an inescapable part of the human condition that we must live with uncertainty. No amount of factual knowledge—though knowledge is essential—can fully prepare an individual or a nation for an unprecedented event. No weighing of the global balance of power is precise enough to prevent miscalculation, blunder, or apparent irrationality. No degree of planning can possibly anticipate all contingencies, eliminate all obstacles, obviate all crises. There is no such thing as the foreseeable future.

To one degree or another, U.S. interests in global natural resources all hang on the slender thread of the unknowable. At this writing, Iran and Iraq—both members of the Organization of Petroleum Exporting Countries—have been at war for years with no end in sight, a conflict that all too easily could escalate to involve enough of the oil production facilities of both countries to dramatically change the face of the international oil market once again. Southern Africa could erupt in political and even military change and conflict sufficient to take certain so-called strategic minerals off the international market for months, perhaps longer. The heat and drought of the summer of 1983 in the United States could visit the country again in 1984 and successive years. Given America's importance in world food trade, depressed production in the United States coupled with poor harvests elsewhere could lead to worldwide hardship and deprivation.

None of these or many other easily imagined nightmares is certain or even likely, but all are plausible. To expect none of them to happen is foolhardy. Yes, any such occurrence would be "abnormal," but normal cannot be defined as ideal. In Bundy's apt phrase, we must recognize the inevitability of the unexpected.

The key quality for Americans, then, as we approach the end of the twentieth century, is adaptability. Natural resources and their services clearly will become ever more important parts of U.S. foreign policy. The only question is whether they will be *ad hoc,* episodic surprises, as in the past, or whether they will be fully integrated components of foreign policy deliberations. In our minds, of course, there is no question: unless the realities of natural resources are integrated comprehensively into its approaches to the rest of the world, the United States will squander an entire spectrum of opportunities for achieving its foreign policy objectives and, indeed, for advancing the cause of pcace.

chapter two

Energy Is an International Good

Milton Russell,
with Mary Beth Zimmerman

Energy is international. The geography of consumption overlaps only imperfectly that of production. International transfers of energy commodities—primarily oil and coal, but increasingly natural gas and electricity—are everywhere essential to prosperity and economic growth. The value of energy trade almost equals that of food and raw materials taken together.[1] Energy pollutants know no political boundaries, and even the global climate may be affected by energy activities that lead to deforestation or releases of carbon dioxide. Improved technologies for winning fuels from the earth or for collecting and converting renewable energy sources to useful work are available worldwide. The financial flows that accompany the transfers of fuels form a major part of the activities of the international financial community.

Whether despite or because of these links, energy has been among the most rancorous topics of the past decade, with problems arising over such diverse issues as nuclear proliferation, the Soviet natural gas pipeline, and acid rain. But the most spectacular set of problems has at its center oil prices and availability, and hence this chapter concentrates heavily on the oil market.

Why oil? True, the surplus in deliverability—the "glut"—moved oil off the front pages in 1982, but the continuing strains wrought by the energy price increases of the 1970s, plausible threats to economic stability of further oil-related developments, and geopolitical problems in the Middle East combine to keep oil on the agenda for national and international action. Moreover, the international energy story is largely an oil story. A single oil market exists from which all consumers must draw and into which all producers must sell. To a first approximation, oil from that market is the incremental source of energy in the world, and oil in world trade provides that increment. Thus, directly or indirectly, energy developments in one country feed into the terms, including the price, at which others produce and consume energy. Alcohol in Brazil, Australian coal shipped to Japan, and added insulation in America's roofs all limit Organization of Petroleum Exporting Countries (OPEC) oil sales, just as natural gas price controls in the United States increase them.[2] Each importer has an interest in having others produce more oil and consume less. Exporters have contrary goals. This means that even domestic energy policy always has an international dimension.

Consensus is lacking in the United States on what that policy should be. Put at its most basic, the issue is to what extent oil markets freed of price and allocation controls can and will lead to the best of possible worlds, and to what extent intervention of various kinds may be desirable. The range of issues is wide: it includes the question of special support to developing countries, for example, in promoting petroleum and other indigenous energy supplies; agreements with other oil importers to coordinate preparation for and response to a future oil supply disruption; and actions to be taken now to shockproof the U.S. economy and those to be undertaken later if another disruption comes. In a sense, we are just now seeing the full effects on the international order of the twin oil price jumps of 1973–74 and of 1979–80. This time of relative tranquility may be used to consider what long-term adjustments may be required in their wake.

Before moving forward to this assessment, it is important to make sure that the framework for considering the problems of the future matches reality. This means taking account of the redistribution of wealth that flowed from the price increases, the effect of those increases on economic activity and prospective economic growth, and the change in the prospective patterns of energy services and uses that have followed. On this basis a number of continuing tensions can be identified, of which the most critical may be the threat posed by uncertain oil supplies.

The Transfer of Wealth

Led by oil, the energy price increases of the 1970s have resulted in the largest peaceful redistribution of wealth and income from a single cause in recent times.[3] Wealth and income have shifted among nations, among regions within nations, among sectors within economies, among occupations, among income classes and, perhaps, among generations.

It also has raised some nations to new prominence and shifted the focus of world attention. If it were not for oil, Saudi Arabia would not now have a nonrotating seat at the International Monetary Fund, much less world attention would be paid to the war between Iran and Iraq, and few would be able to cope with Abu Dhabi in a crossword puzzle.

To most people, the massive changes our societies have experienced have had no discernible causes: they have not come because of war and conquest, technological breakthroughs, or natural disaster. The amount of oil in the ground did not suddenly shrink; oil exporters and domestic producers did not suddenly grow smarter or more productive; long-delayed results of forward-thinking investment did not suddenly begin to pay off. Energy consumers became poorer, but not because they became dramatically more lazy, or that things they produced overnight became less attractive on the world market.

Instead, the perceived *future* value of oil in the ground increased radically. This came as the oil-producing countries, rather than companies, gained control over reserves in the Middle East, then the incremental source of world energy supply. The increased security of the right to produce the oil sharply reduced the discount rate from that appropriate to a foreign company subject to expropriation.[4] Hence, the willingness to produce oil fell, and its price rose rapidly, in part just to catch up to the level that more secure property rights might have dictated before. In addition, a small group of oil-exporting countries realized their mutual advantage in seeking to avoid competition for world markets, and acted simultaneously to restrain supply and further push up price. Other energy prices followed oil upward.

The Oil Shocks

The worldwide economic boom of 1972–74 led to intense pressure on oil supplies, a trend that carried forward through political events that temporarily shocked the oil supply system, first in 1973–74 and then in 1979–80. The restriction of supply after the late 1973 Arab oil embargo was the impetus to 1974's increase in oil prices (table 2-1). Ever since, the Arab oil producers have been targets for the hostility of those who saw no other reason for the increase in the price of oil. (Consider that people otherwise sensitive to racial or ethnic slurs seem to find acceptable ads or cartoons featuring a rapacious Arab in a burnoose.) Along with its speed and size, the perception of the energy price revolution as exploitative is important in explaining the divisiveness and tension it brought to international relations. The fact that the 1979–80 price increase can be tied directly to the revolution in Iran did nothing to dispel that image. That Iranians are not Arabs and that the immediate effect on Iran of drastic cuts in oil exports was negative probably were lost distinctions to most Americans.

Faced with these events, the natural reaction of importers was shock at the size of the change and anger that it had occurred. In most industrialized countries, that anger has been directed at the oil-exporting nations, with some of it aimed at such scapegoats as the oil companies. Major consuming nations—the United States principal among them—which are presumed to have failed to exercise enough control over their oil appetites also have come in for criticism. In the Third World, the anger, but not the distress, has been diluted by admiration for the daring of the oil exporters and a feeling of solidarity with them. The oil exporters themselves, after all, are underdeveloped.

By and large the industrialized nations see a picture of exploitative, devastating price increases, new (undeserved?) wealth, and increased poverty. But this picture appears very different to the oil-exporting nations. They see how much they were cheated by the *depressed* price of oil until 1973, how poor they are now, and how poor they will be in the future when the oil runs out. Besides, they say, the shift is exaggerated. Oil still is a small part of the world economy: total oil

Table 2-1. Crude Oil Prices; Nominal and as Adjusted by the U.S. Gross National Product and OPEC Import Price Deflators, 1972–82

Price	*1970*	*1971*	*1972*	*1973*	*1974*	*1975*	*1976*	*1977*	*1978*	*1979*	*1980*	*1981*	*1982*
Nominal price of OPEC crude (current U.S.$, official f.o.b.)[a]	1.35	1.75	1.90	2.64	9.56	10.46	11.51	12.88	12.93	18.67	30.87	34.50	33.11 E
Deflated U.S. price (1973 U.S.$)[b]	1.56	1.93	2.01	2.64	8.79	8.80	9.20	9.73	9.09	12.08	18.28	18.66	16.89 E
Index of deflated U.S. price (1973 = 100)	59	73	76	100	333	333	348	369	344	458	692	707	640 E
Deflated OPEC export price (U.S. 1973 $)[c]	1.71	2.22	2.41	2.64	8.03	7.92	9.06	9.27	7.98	10.20	14.84	17.25	17.34 E
Index of deflated OPEC export price (1973 = 100)	65	84	91	100	304	300	343	351	302	386	562	653	657 E
U.S. implicit price deflator[d]	86.48	90.79	94.56	100.00	108.82	119.00	125.10	132.40	142.20	154.50	168.90	184.90	196.00 E
OPEC import basket price deflator[e]	———	79	———	100	119	132	127	139	162	183	208	200	191 (July)
U.S. $ value of SDR[f]	—	1.00	1.09	1.19	1.20	1.21	1.15	1.17	1.25	1.29	1.30	1.18	1.12 (2nd quarter)

Note: E = estimate.

Sources: [a] Price reported is for Saudi light prior to 1977 and is an official average price weighted by export volume thereafter; 1970–76 prices are from *Petroleum Intelligence Weekly*, April 22, 1982, p. 11; 1977–81 prices are from U.S. Central Intelligence Agency, *Economic and Energy Indicators*, March 19, 1982, p. 10. The 1982 price was estimated from *Petroleum Intelligence Weekly*, October 4, 1982, p. 2, and December 6, 1982, special supplement. The price of Saudi light was a good indicator of the crude market until the split price pattern emerged with the December 1976 OPEC Doha decision. Prices shown here provide a reasonable picture of broad movements; shifts in the ancillary terms of the bargain are hidden, however, and it is therefore difficult to compare pre- and post-1974 prices.

[b] Nominal crude price deflated by U.S. GNP implicit price deflator.

[c] Nominal crude price deflated by the OPEC weighted imported basket price deflator.

[d] Executive Office of the President, *Economic Report of the President*, February 1983, p. 236.

[e] Morgan Guaranty Trust, *World Financial Markets*, 1970–73 from May 1979, p. 3; 1973–80 from December 1980, p. 8; 1981–82 from September 1982, p. 8.

[f] International Monetary Fund, *1982 International Financial Statistics* vol. 35, no. 9 (September) p. 14. The SDR or Special Drawing Right is an international reserve asset created by the Fund.

sales of all OPEC countries—with a third of a billion people—in 1981 were equal to only about five weeks' worth of U.S. gross national product that year, and for 1982 were down to three weeks' worth (see table 2-2).[5] This mismatch of perceptions poisons cooperation between importers and exporters on energy matters. And the real story, properly interpreted, is less traumatic than it is perceived to be based on casual observation.

Perception versus reality. Careful consideration of what really has happened thus is important. Consider first the price of oil compared to the cost of things oil-exporting countries buy—a measure of what exporters put into the world economy compared to what they take out. On this measure, the average price for 1974 was three times that of 1973—not the 3⅔ times (or "quadrupling") measured in nominal terms and now almost legend. (See table 2-1 for price trends and data based on alternative concepts of what is important.) Prices then fluctuated around this level through 1979, when the second shock boosted prices again to where, by 1981, the price was 6½ times that of 1973 in terms of what OPEC exporters buy. Through the end of 1982 the real price on this measure continued to creep upward because of the recent strength of the U.S. dollar, despite the stability or decline in the nominal price.[6]

The picture from the consumer perspective is somewhat different. Changes in the international exchange value of the dollar are not relevant to the United States, of course, so what matters are changes in domestic prices. Adjusted by the GNP implicit deflator, the 1982 oil price level was 6½ times that of 1973—and not, as often reported, the 12½ times based on nominal prices (see table 2-1). The experience of other countries was diverse, affected by the value of the local currency in dollars as well as by internal inflation rates.[7] The decline in the value of the dollar through most of the 1970s meant that the oil price increase was cushioned for consumers in countries such as Japan, Germany, and France, but the subsequent rise in the dollar led to substantial real oil price increases, even during a period when the nominal price fell (table 2-3). For example, the real price of oil to Japanese consumers *fell* nearly 30 percent between 1974 and 1978 while it *rose* 2 percent in the United States and 26 percent in India. Between January 1981 and September 1982, there was a 15 percent *decline* for the United States and a 17 percent *increase* for France.

Table 2-2. Revenues and Current Account Surplus in Nominal Dollars and Adjusted for OPEC Import Prices, 1972–82

	1972	*1973*	*1974*	*1975*	*1976*	*1977*	*1978*	*1979*	*1980*	*1981*	*1982*
Estimated oil revenues (billions U.S.$)[a]	14.4	23.1	91.1	94.7	109.3	128.9	119.2	199.0	274.0	255.0	200.0 E[b]
Revenues deflated by OPEC import price deflator (billions 1973 U.S.$)	n.a.	23.1	76.6	71.7	86.1	92.7	73.6	108.7	131.7	127.5	104.7 E
Current account surplus (billions U.S.$)[c]	n.a.	5.0	64.2	29.7	34.4	24.3	0.6	62.0	105.0	59.0	(9.5) E
Surplus deflated by OPEC import price deflator (billions 1973 U.S.$)	n.a.	5.0	53.9	22.5	27.1	17.5	0.4	33.9	50.5	29.5	(5.0) E
OPEC implicit import price deflator[d]	n.a.	100.0	119.0	132.0	127.0	139.0	162.0	183.0	208.0	200.0	191.0 (July)

Note: E, estimate; n.a., data not available.

Source: [a] Shell Briefing Service, Shell International: 1972 from "Oil and Gas in 1977," June 1978, p. 9; 1973–78, "Oil and Gas in 1978," May 1979, p. 9; 1979 revenues from *Petroleum Economist,* June 1980; 1980 from Morgan Guaranty Trust, *World Financial Markets,* March 1982, p. 4; 1981 from Morgan Guaranty Trust, *World Financial Markets* (September) 1982, p. 8.

[b] Revised from Morgan Guaranty Trust, *World Financial Markets* figure, reported September 1982, to account for decline in assumed oil production in last quarter.

[c] Figures for 1973–79 from Morgan Guaranty Trust, *World Financial Markets,* (March) 1980, p. 3; 1980, from March 1982, p. 4; 1981, from September 1982, p. 8; 1982 from *OPEC Bulletin* vol. 13, no. 7 (August) 1982, p. 11.

[d] Figures for 1972–73 from Morgan Guaranty Trust, *World Financial Markets* May 1979, p. 3; 1973–80, from December 1980, p. 8; 1981–82, from September 1982, p. 8.

While price is important, oil revenues and the current account surplus or deficit of OPEC are truer measures of trends and gyrations in energy markets. These include the effects of changes in quantities sold. Adjusted for what revenues can buy of what exporters want to consume, revenues were 3⅓ times as high for 1974 as for 1973, and 5½ times by 1981 (see figure 2-1 and table 2-2). With the decline in oil sales, the quantity of goods OPEC could purchase fell in 1982, even though a strengthening of the U.S. dollar resulted in terms-of-trade benefits that more than offset declining nominal prices.[8] Real revenues were down some 18 percent from 1981. The OPEC current account surplus shows the combined effects of revenue changes and of growth in OPEC purchases. Because purchases lag revenues, the surplus in real terms went up almost eleven times in 1974 from 1973, fell almost to zero by 1978, was back near its 1974 peak in 1980, and in 1982 showed a *deficit* of about $10 billion. Of course, the OPEC accounts taken together mask vast differences among the thirteen OPEC countries. Many of them have run deficits most years since 1973.

Exporters are poor. Comparisons over time are interesting, but recognition that oil-exporting countries, taken together, are poor is more instructive, in terms of prospective developments. Per capita incomes of OPEC countries still are a small fraction of those of the industrialized world—perhaps one-sixth or one-seventh; they were one-thirteenth in 1971. True, some thinly populated countries are very rich: Kuwait, the United Arab Emirates, and Qatar have among the highest per capita incomes in the world. But even when the more populous countries of Indonesia and Nigeria—poor in any terms—are taken out, OPEC countries have per capita incomes about half that of the industrialized world average. Saudi Arabia had a per capita income about that of the United States in 1980, the last year for which data are available, and now (1983) probably has fallen behind because of the decline in oil sales and prices.[9] Clearly, oil makes one richer than being without oil, but oil does not make a country rich.

Moreover, a country rich in current income still can be poor and underdeveloped. Industrialized countries have masses of capital—the infrastructure of transportation networks, acquired skills, trained workers and public works—that yield current services but do not show up in any set of social accounts. The current income of newly rich nations is poured into building that which our forebears left to us.

Table 2-3. Changes in Effective Price of Imported Oil, by Country, for Selected Periods, 1973–1982

	1973–74 262% Nominal price increase			*1974–78 35.3% Nominal price increase*		
Country	*Percentage change in real price*	*Change in exchange rate*	*Change in general price level*	*Percentage change in real price*	*Change in exchange rate*	*Change in general price level*
Brazil	214	10.8	27.6	−1	166.1	264.8
France	244	8.1	13.7	−13	−6.2	46.1
Germany	228	−3.0	7.0	−11	−22.4	17.7
India	197	4.7	27.8	26	1.1	8.2
Italy	239	11.5	19.1	−2	30.5	79.3
Jamaica	231	13.6	24.4	11	57.9	93.3
Japan	213	7.5	24.3	−29	−27.8	37.1
Kenya	213	2.0	17.8	−18	8.3	78.3
Korea	198	1.9	23.8	−12	19.2	83.7
Philippines	171	0.4	34.2	10	8.6	33.2
Portugal	184	3.0	31.5	10	72.9	117.4
Thailand	190	−1.2	23.2	6	−0.2	27.0
Turkey	207	−1.6	15.9	−9	74.3	158.3
United States	226	n.a.	10.9	2	n.a.	32.3

Note: The percentage change in the real price of oil was calculated as $R = [(1+N)(1+E)/(1+I)] - 1$, where N is the change in the nominal price, E is the change in the exchange rate, and I is the change in general prices; n.a. = not available.

Sources: 1973–78 adapted from Michael J. Coda and John E. Jankowski, Jr., "The Real Price of Imported Oil Revisited," *Energy Journal*, October 1983; 1981–82 figures are from the International Monetary Fund, *International Financial Statistics*, December 1982. The change in the nominal price of oil is based on the price series in table 2-1. Exchange rates and general prices are from the IMF, *International Financial Statistics*, September and December 1982, September 1981, and November 1977.

And oil exporters have another argument. Even if for the time being oil prices rise substantially faster than the costs of things they buy, and oil revenues more than keep pace with inflation, they see depletion of oil as the current expenditure of their people's future. It is the current *income* of industrialized countries that is being exchanged for the nonreproducible *wealth* of the oil exporters. As Saudi Arabia has oil, the United States has its vast, fertile grain belt. But the oil exchanged for food is gone forever while, with proper care, the productivity of the soil remains. For exporters, unless oil wealth can be transformed into a permanent income stream through investment, the endowment of oil means only a brief (and in some cases limited)

1978–80 139% Nominal price increase			*January 1981–July 1982 5.0% Nominal price decline*		
Percentage change in real price	*Change in exchange rate*	*Change in general price level*	*Percentage change in real price*	*Change in exchange rate*	*Change in general price level*
150	191.7	179.1	−14	163.3	189.7
79	−6.2	25.5	17	47.5	19.9
97	−9.4	9.8	7	22.7	9.3
94	−4.0	18.5	−2	20.1	16.3
73	0.9	39.1	10	44.9	25.6
81	24.2	63.9	−13	0.0	9.0
130	7.7	11.9	16	26.3	3.7
87	−4.0	22.9	9	43.6	25.6
97	25.5	52.1	−7	11.3	14.0
74	1.9	40.0	−9	11.2	16.2
89	13.9	44.4	8	56.7	37.6
84	0.7	30.5	−5	11.3	11.4
125	213.3	233.4	11	84.1	56.9
89	n.a.	26.3	−15	n.a.	12.2

interlude of greater prosperity, in a past—and a future—of want and relative deprivation. What importers see as unjustified windfalls, exporters, with equal fervor, perceive as their (vanishing) birthright.

Thus, the price shocks that left oil importers sorely aggrieved are seen by exporters as relatively moderate and surely justified. These opposite perceptions contribute to conflict and threaten to poison importer-exporter cooperation that could help promote future oil market stability.

Effects on the Economy

The two oil price shocks have done more than redistribute income and wealth. They also have led to lower present and prospective economic output, and to particularly severe consequences for the developing countries.

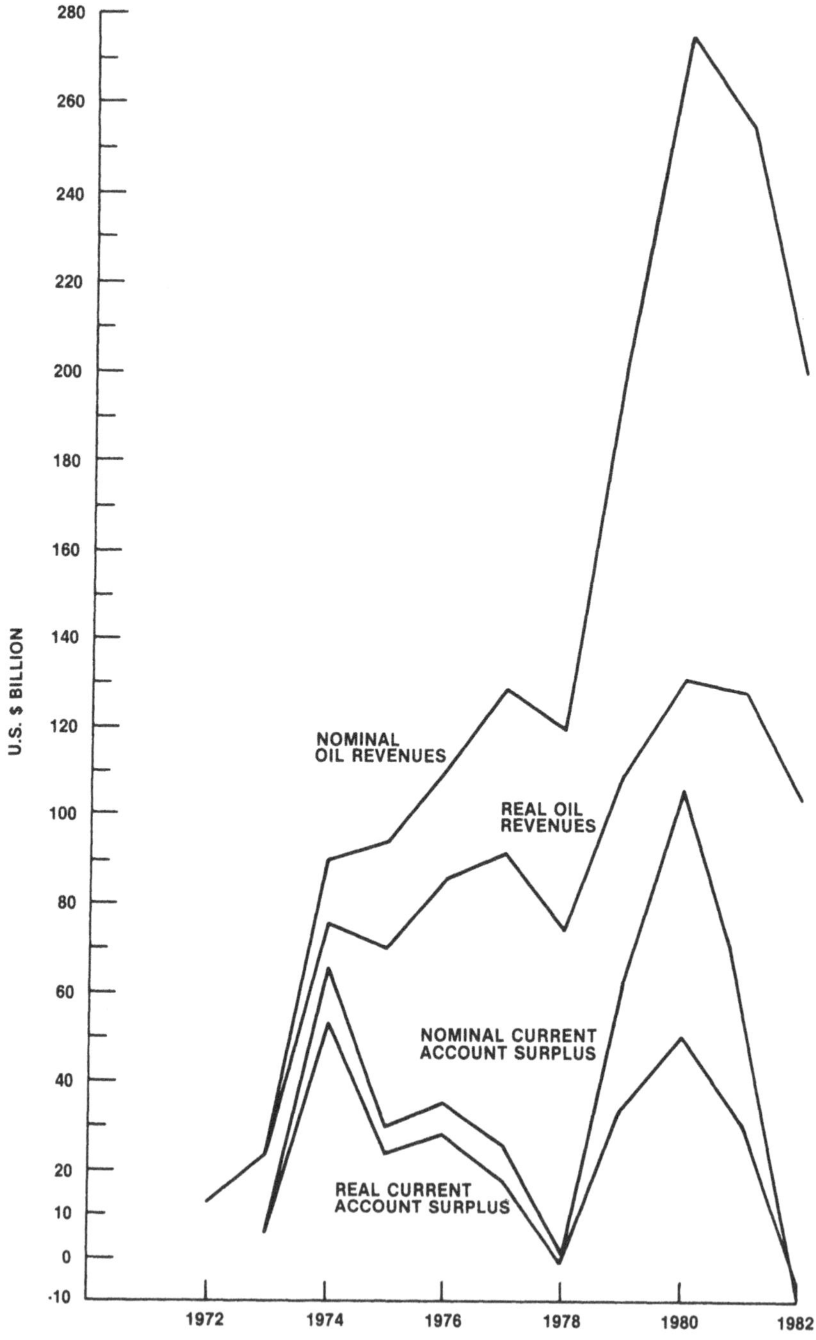

Figure 2-1. OPEC oil revenues and current account surplus, real and nominal. Revenues and current account surplus are deflated by the OPEC implicit import price deflator. Based on data from table 2-2.

Substantial though the gap is, the loss to the world from increasing oil prices in two leaps—compared to a more gentle increase consistent with requirements to meet demand and to allocate oil efficiently over time—cannot be quantified precisely. Current consumption has been sacrificed, the growth path of future income has been permanently lowered and, probably at least for the rest of this century, its trajectory depressed. What has occurred is a loss to the world as a whole: gains by exporters are far overshadowed by losses to others.

The reasons that the loss to oil importers was greater than the gain to exporters are reasonably straightforward. Most important, the shock of the first price increase brought recession and inflation, which led to lower world output. Recovery from the 1974–76 recession never was robust and, in any event, was cut short by the oil price spiral that began again in 1979. The world economy has yet to recover. While the economic system is far too complex to warrant ascribing specific causes for its movements, surely the two oil shocks bear major responsibility for the overall poor macroeconomic performance of the past decade. Output has suffered, too, because of the temporary—in some cases, permanent—underemployment of other factors, including specialized labor, caused by the rapid change in the relative price of energy. Importers also have been driven to make investments in conservation, alternative energy production, and early replacement of energy-using capital that have absorbed resources that otherwise would have been available to do other things with higher payoffs.

Impact on the Poor

Of special interest is the effect of the oil price increases on the poorer countries of the world. The developed countries also received a smaller share of a smaller world pie as oil exporters took more but, taking the long view, the effect on the developing countries probably was more severe. For one thing, the first shock came at a time when oil consumption in developing countries was rising rapidly—at perhaps 10 percent per year for the pre-1973 decade—to fuel economic growth, to support the change from traditional to modern living and working patterns, and to relieve the pressure on traditional fuels.[10] In addition to raising fuel bills, the price increase gave a second jolt to international payments difficulties because it depressed markets in the industrialized countries. Industrialized countries recouped some of their lost markets as exports rose to the oil-exporting economies, but this outlet was mostly denied the poorer countries who had little to sell that the oil exporters wanted.[11] Further, the developing countries were severely affected because they are poor. The level of living of those at the margin of subsistence was reduced, which is a matter different in kind than the declines in levels of affluence experienced in the developed nations. The loss was not made up in aid; the 1979–80 oil price increase added over 1½ times as much to the 1981 oil-import bills of the net oil-importing developing countries as all the development assistance they received that year.[12] Over the longer term, the annual oil import bill of net oil-importing developing countries (in 1982 dollars) went from about $9.5 billion in 1973 to about $71 billion in 1981.

The annual $60-odd billion added by 1981 can be compared to the $30 to $40 billion annual official development aid in 1974–80.[13] These numbers are not strictly comparable, of course, but the point is clear: the oil shock has overwhelmed the aid efforts of the affluent world, notwithstanding recent generous contributions by oil exporters.[14]

Until recently, though, many developing countries, especially those at the upper end of the income scale, were spared immediate severe sacrifices. Oil exporters were unable to consume or invest at home all their new wealth, forming a ready pool of world savings from the consumption denied developed countries by the rising cost of oil. Through loans, some of these resources found their way into the developing world. Some countries thus were able to maintain consumption, but only at the cost of surrendering a larger portion of their future growth dividend to others when the debts come due. But this recycling seems to have run its course. Oil surpluses no longer exist—the OPEC countries ran a current account deficit of some $10 billion in 1982—and the future growth of the developing world has mostly already been swallowed up to service existing debt obligations. The roll call of developing countries at or beyond borrowing limits and now rescheduling debts seems to mount daily (as does concern about the liquidity of the international financial system). As the commercial credit flow dries up and concessional aid gets more scarce, more and more countries must restrict either current consumption or current investment. Political stability, economic imperatives, and humanitarian concerns limit the ability of poor countries to restrict consumption. As investment inevitably shrinks, future growth rates will be depressed.

Taking the world economy as a whole, however, the system so far has proved resilient. Recessions have not cascaded into panics and worldwide depressions of sorts known before. Timely changes have allowed the financial system to function reasonably well, albeit with creaks and groans, though additional remedial action now may be required. In short, the apocalyptic, energy-centered economic crisis, heralded nine years ago and often since seen on the horizon, has not materialized—at least not yet. No one would suggest that the system even now is not under severe strains, and further tests will come. But so far, the outcome has not been collapse but rather a process of reasonably successful adjustment, however grinding and painful it seems along the way.

Effects on the Energy System

Before 1973, oil was everywhere a cheap, convenient, and reliable energy source. Oil still is convenient, but it no longer is cheap or reliable, and less of it is used as a consequence.

Consider the effects of the price increase. It slowed economic growth all over the world, meaning less energy demand. It reversed the flight from coal and from such renewable fuels as wood and dung. It hastened the development of natural gas resources that before had been little exploited outside the United States. It stimulated earlier research and development on any number of energy technologies, including synthetics from coal and shale. It focused new attention on ways to save energy, and made the labor and capital required to achieve these savings worth expending. While the evidence is difficult to interpret, the Btu/dollar ratio has dropped about 18 percent in the United States since 1973.[15] In all these dimensions, moreover, the impact of the price changes has not yet been spent; it takes a long time fully to turn the energy system around.

The implications are clear: compared to a no-shock trajectory, less energy, less oil, but more of other energy sources will be used between now and the year 2000; in 2000 less oil will be demanded (because of more time to learn to conserve and to use other energy sources), but more oil will be left in the ground to use; and the major transition to alternatives for things oil does best may be more gradual and thus less traumatic than it would have been had we not borne the shock in the 1970s. As an indication of the extent of the changes, consider that the conventional wisdom in the early 1970s was that free world oil consumption would be about 80 to 90 million barrels per day by 1985. By 1975, the estimate for 1990, five years later, was down to 70 to 80 million barrels per day. From 1983's vantage point, it seems likely that fewer than 50 million barrels per day will be used at the end of this decade.[16] Figure 2-2 indicates the way production estimates have been changing as analysts have observed and come to understand the magnitude of the shifts that have occurred.

These changes in oil production will be mirrored in the path of oil prices compared with the scenario of no oil shocks in the 1970s. Prices have risen faster. From current levels, then, the rate of future oil price increases originating in underlying economic phenomena (as contrasted to possible exogenous forces such as wars that destroy producing regions) will be slower than could have been expected. Moreover, the ultimate level of oil prices likely will be lower. It may be cold comfort, but because this generation pays more, the next may pay less. But note: the world as a whole surely would have been better off with steadily rising prices than with the shocks that gave us this more rapid transition. However, all the pain was not suffered in vain.

Production and Trade Changes

Not only did the oil price shocks encourage conservation and promote use of alternatives, but also they changed patterns of oil production and trade. While helpful now, these shifts may not bode well for stable oil markets in the future.

To understand these changes, consider that marginal oil production costs—that is, the cost of producing an additional barrel—are lowest by far in the Middle East. But with prices at current levels, rather than getting more from places like Saudi Arabia, where the average well produced about 15,000 barrels per day in 1981 and costs were at most a dollar or so, about 50,000 oil-related wells were drilled in the United States, where the average well produced about 16 barrels per day.[17] The industry explores in the incredibly expensive far Arctic, in the deep waters of the Baltimore Canyon, in the stormy North Sea, and in the rain forests of Peru. It uses up resources with current alternative uses to find and produce oil elsewhere rather than to

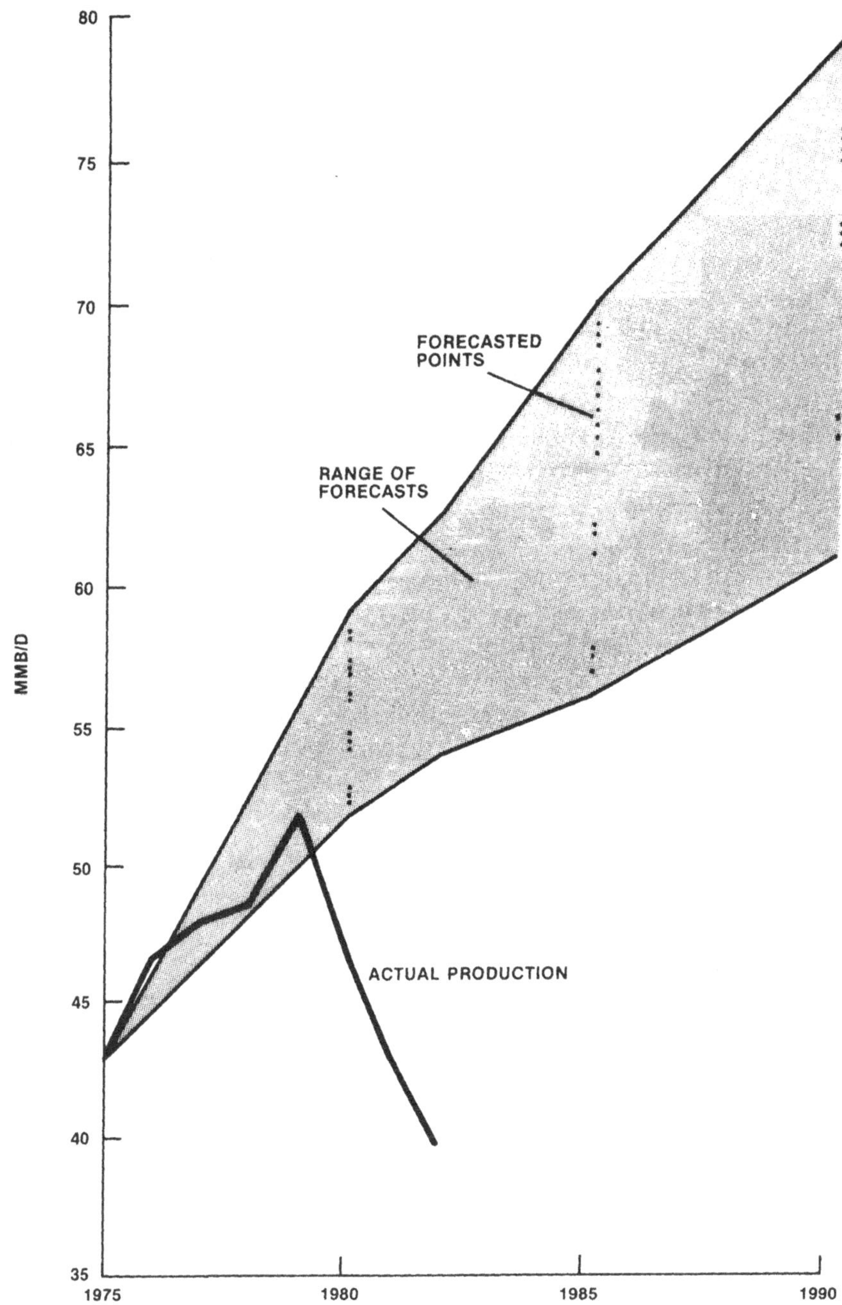

Figure 2-2. Free world oil production forecasts and actual oil production for 1975–78. Production projections are adopted from U.S. Senate Committee on Energy and Natural Resources, *Energy: An Uncertain Future,* December 1978, fig. 12, p. 40; actual production figures are from *Petroleum Economist* (December) 1982, p. 524. For a list of the projection studies included, see *Energy: An Uncertain Future.*

consume more of the oil that flows almost without such costs out of the Middle East. In other words, as *prices* of low-cost oil owned by major exporters have risen, others have expended greater *costs* to produce their own. This does not make sense from a global economic perspective, but it does from that of other producers, as long as their real costs are less than world prices. Thus, for another decade or so, until the inventory of economic drilling prospects at current prices is more depleted, the price increases of the last decade mean proportionately more oil produced within oil-importing countries, less oil traded internationally and, of that traded, more supplied by new exporters such as Mexico.

The impact of higher oil prices on production patterns already is dramatic. OPEC production for 1982 averaged approximately 18.5 million barrels per day, down from more than 31 million barrels per day in 1977; excepting the centrally planned economies, this represents a decline from 64 percent of total world production to 46 percent.[18] OPEC producers are at the end of the line. Oil is the residual claimant for energy demand, and OPEC producers are the residual suppliers of the world's oil. As a residual of a residual, OPEC production will gyrate; a 1 percent change in non-OPEC energy use may change OPEC exports by as much as 4 to 6 percent.[19] Since 1980 this arithmetic has led to a decline in OPEC sales but, as noted later, the lever works as well in the opposite direction.

While reduced dependence on some of the more volatile areas of the world has added stability to the world oil market, today's gains may be coming at a cost to be paid in the future. As the oil resource base outside the Middle East is depleted more rapidly, less dependence today is being exchanged for greater concentration of oil supplies in the hands of a few suppliers in the future. Discoveries outside the Middle East large enough to change this picture materially seem unlikely, as does development of synthetic fuels in large volumes or other replacements for liquid fuels, especially for transportation.[20] Even though too much can be made of these facts—for example, over the coming decades oil likely will make up a progressively smaller proportion of the world's total energy supply as coal, nuclear power, renewable energy sources, and synthetics take on more of the total load—based on current evidence, conventional oil will remain important and oil vulnerability will remain a concern through this century and beyond.

Energy Problems and the Role of Government

What energy matters remain on the policy agenda? The oil shocks of the 1970s presented four classes of problems.

"Fairness." The first was to deal with the change in real incomes within consuming countries. Some were made poorer and others richer. Societies had to decide whether their income distribution after the energy shocks continued to meet standards of equity or whether new taxes or welfare schemes were justified, especially for the period during which longer-term adjustments were in process.[21]

Minimizing costs. The second class of problems was to adjust behavior and the capital stock to minimize the avoidable costs imposed by higher oil prices. This task required little government intervention, though in practice it received much: regulations, subsidies, and exhortation abounded. To the extent that some interventions occurred, others seemed necessary to offset the effects of the first. For example, because oil prices were held below world levels, subsidies to alternative energy sources and to conservation seemed to make sense. As experience after the price increases grew, however, evidence mounted about the adaptive power of consumer and producer choices when they were free to respond to correct price signals. Ushered in with a phalanx of government interventions, the process of changing the economy to conform to the higher cost of energy yielded a real-world experiment that attested to the efficiency of a free market system, especially in the case of large and discontinuous changes.

Long-term transition. The third problem was to adjust the level and pattern of investment going into the long-term transition toward renewable or less-limited energy sources because of new information on the level and prospective path of oil and other energy costs, of economic growth, and of energy consumption. Inherently, some aspects of long-term transitions are not handled well by the private sector: (frequently expensive) basic research is required, private and public discount rates may differ, benefits and costs of long-term investments are hard to internalize, and risk-return-scale relationships sometimes argue against private activity. Moreover, the problem is global: the peoples of the world share a common resource base and can call on common technology to exploit it. Thus, elements of preparing for the long-term transition are international in character. On familiar grounds of market failure, then, national governments and international organizations need to be involved in preparing for future patterns of energy production and use. But nothing about the oil price shocks altered this situation. They simply brought greater awareness of the need to plan ahead and, for a while, greater willingness to act. Indeed, as suggested above, the increase in oil prices stretched out the availability of conventional energy sources and hastened the introduction of alternatives. Consequently, other things being equal, optimal levels of social investment in the energy future fell, not rose, with the extraordinary price increases of the 1970s.

Vulnerability. The final problem was—and is—oil vulnerability. U.S. national security interests were not tied to oil before 1973, at least not to the same degree. But the Arab oil embargo and OPEC's first price hike confronted the United States with a set of conditions that allowed discontinuous shifts in the price and availability of oil that would threaten national security, damage the domestic economy, and disrupt the world economic system. New roles for government were created to meet these new challenges.

What changed? Put most simply, by the early 1970s the marginal source of oil to serve as the buffer against exogenous changes in demand or supply shifted from secure suppliers in the United States to unstable ones in the Middle East.[22] This shift, especially threatening in the context of rapidly rising total oil imports relative to consumption, meant a change from control by somewhat decentralized decision makers (oil companies, the Texas Railroad Commission, and federal oil import control authorities) whose actions were constrained by some identification with consumer interests, to control by a few producing countries for which political goals sometimes vied with the desire to maximize revenues over the shorter or longer run. With swing production now concentrated at a few locations in the Middle East, there were risks unknown before: wars in that troubled region, terrorism or revolution within oil-producing states, political adventurism with oil as the weapon, and major accidents or acts of God now could interrupt oil supplies. The shift of control to sovereign governments meant that levels of oil production and hence prices could be linked to a wide range of events and motivations far less predictable than maximizing producer revenues, thus adding more uncertainties. Finally, opportunities for economically advantageous price manipulation arose when a few governments controlled both marginal supplies and a large share of total output. The number of agents required to collude is small, no legal recourse exists against sovereign governments, and the benefits to them of higher prices may offset the loss of sales from restricting production. The converse also holds. Preemptive temporary price reductions to discourage investment in alternative production became a viable strategy whose threat alone helped sustain a long-term market for their oil.

This shift opened a new arena for great power conflict as well. The Soviet presence in the Middle East is dictated by proximity and expressed through client states and friendly political movements. But more than territory is involved. The Middle East is strategic because of oil, and oil is important because of the effect changes in its price and availability have on economies. It is in the obvious interest of the West to deny the Soviet Union dominance in the region, both to preclude physical access to its oil wealth *and* to prevent the USSR from setting the conditions on which the United States and its allies get fuel. The matter goes well beyond geography or military security. On a global scale, it has much to do with economic strength and vulnerability and the balance of economic advantage that can be transferred into military strength and diplomatic influence.

One policy conclusion that follows is that diplomatic and strategic—even military—action and preparation may be required to limit the probability, size, and duration of any disruption, and to limit any potential direct Soviet threat. These responsibilities inherently are those of government; in this sense, new roles were created by the changed oil market of the early 1970s. But a role for government exists in addition to the diplomatic, strategic, and military dimensions. Since it is the economic cost of disruptions that drives U.S. concerns and those of its allies, limiting economic vulnerability can both substitute for and complement other approaches. The question thus comes down to whether government can take any actions in the economic domain that effectively would limit vulnerability to oil disruptions.

The Continuing Challenge of Energy Security

To some degree, increased vulnerability of oil supplies and hence uncertainties about oil price and vulnerability are internalized in the decisions made by private parties, and cost-minimizing adjustments are made. The market works here, too. But, because of spillover effects discussed below, such actions inherently are inadequate; action by national governments, by importing governments collectively, or even by importing and exporting governments working together is required if avoidable costs are to be minimized. The new conditions under which marginal supplies of oil are produced in principle as well as practice have enlarged the role of government and provide a serious challenge for energy policy for the 1980s and beyond.

However, the willingness to act faded with falling prices and the roughly 13 million barrels per day of global excess capacity that greeted 1983.[23] Scenarios that would yield supply reductions sufficient to cause difficulty required much less plausible combinations of circumstances than at any time since the early 1970s. The price shock of 1979–80 had, for the time being, built up resistance to further shocks. But the protection was neither absolute nor permanent.

The system could be stretched too far. The excess capacity on which the optimistic view depends is concentrated in the very region most vulnerable to disruption. If supplies are lost there, so also is the ability to make them up. The margin of safety that seems to exist is partly illusory. And the matter could grow worse over time. As of 1983, private stocks have been drawn down to near-minimum levels. Hence, reliance must rest on surges of production to meet extra demand or to replace disrupted supplies. However, production capacity responds to output levels. Exporters are loathe to invest in new facilities when those they have are underutilized; thus capacity will fall as reserves are depleted and new ones are not developed. Moreover, oil consumption likely will revive as the effects of the earlier shocks wear off, real price declines occur, economic growth resumes at a higher rate, cyclical downturns are reversed, and inventory reduction no longer is possible. This will shrink the safety margin from the demand side. While no confident, specific quantitative predictions can be made, it is likely that, as the 1980s proceed, oil use will move closer to capacity levels chosen at existing prices, markets will tighten, and scenarios for disruption leading to distress again will become more plausible.[24]

If this picture is close to reality, it suggests that oil vulnerability will remain a problem for the near-to-medium future. Granted that the magnitude of the potential costs is worthy of concern, some government action may be called for if private efforts are inadequate. While specifics of policy are beyond the scope of this chapter, the guiding principles and constraints for domestic and international initiatives can be explored.

Energy Security: The Domestic Challenges

Oil vulnerability poses a peculiarly difficult problem for social policy for two reasons. First, the benefits from acting to reduce vulnerability lie in the future, are uncertain, and are shared—with oil importers abroad and a partially different generation at home. In direct contrast, the costs of preparations are in the present, are certain, and are borne by today's voters. Second, actions of government will affect, often adversely, the preparations taken by private agents. It is particularly hard to design and implement policy where such interactions are strong.[25] Yet, risks remain, and something needs to be done.

Reducing Vulnerability: The Goals of Policy

What are the goals of domestic policy in the light of this threat? Does government have a role in achieving them?

Provide more oil. The first objective is obvious—to provide more oil to the production and consumption process in a disruption. If the market is allowed to work, extra output might come from nondisrupted producers who push existing facilities harder in response to higher prices. Precautionary inventories can be drawn down when higher prices make profits possible, and holding oil more costly. Working stocks can be reduced as it pays to devote more management effort to inventory control. Some of these possibilities are related to actions taken before an interruption. For example, excess production capacity or excess stocks can be maintained that would not pay if the potential for disruption-related higher prices did not exist. Other actions can be taken during an interruption. Relying on private-sector actions means that all parties must have the socially beneficial signals—to build stocks before and release them during a shortage, and to use less oil and produce more oil (and other energy sources) during a supply interruption. The force of these incentives for getting more oil during a disruption depends on private expectations that government will not deflect the price impact of disruptions through such measures as price controls; part of the role of government may be to foreswear some forms of intervention, and to do so credibly.

Even with a noninterventionist policy that disavows price controls and allocations, however, the spillover of benefits to others (as the increase in the world price is dampened by private additions of oil to the market) assures that the private sector will not carry preparations or responses as far as even the sum of private interests requires. Consequently, intervention through a large government-owned or -mandated strategic petroleum reserve, for example, makes sense on market grounds. When the issue is cast into broader terms, the case for stockpiles is enhanced. For example, by relieving the immediacy of the economic strain, stockpile releases can dampen the temptation for extreme measures, provide time for diplomacy to work, and thus lessen the chance of peace-threatening confrontations in the Middle East or elsewhere. Other policies also may be appropriate.[26]

Stabilize the economy. The second goal is to enhance the economy's ability to maintain output and consumption when oil supplies are lost. Two related classes of options exist. Other energy sources such as natural gas or coal can be substituted for oil so that essentially the same production and consumption processes are maintained. Alternatively, production and consumption patterns themselves can be changed, so that the loss of oil has a smaller effect on output and consumption. As above, action must be taken in normal periods if fuel substitution possibilities and oil-flexible production and consumption processes are to be available when they are needed. The private sector will respond to the degree that fluctuations in prices are expected and that profits can be made from preparations for them. But again, some of the benefits of reducing oil use in an emergency are shared with all users while preparation costs are borne alone. Moreover, private parties must take into account the fact that government may short-circuit price signals from supply reductions, as it has in the past. They also must discount the value of their preparations because other policies may be introduced—stockpile drawdowns, for example—that will dampen the price increases required to justify precautionary investments. Thus, there is reason to believe that private incentives will be inadequate for socially optimal preparations.

Allocate by price. A further element is to make sure that the oil that is available is used most productively. While some limited exceptions may possibly be identified, this end can best be accomplished through use of the price system and normal market channels to allocate both regular supplies and any releases from strategic reserves.[27] The problem for policy is to counter the political pressures to allocate oil differently, and to balance sacrifices.

Create credible policy. The realities of the U.S. political system are such that it is unrealistic to expect laissez-faire in a major disruption. In any event, governmental declarations that there will be no intervention simply are not believed and accordingly are dismissed. Private responses consequently take a variety of alternative scenarios into account in planning, leading to inefficiency and probably to underpreparation.[28] Far more effective—because it is more credible—would be an approach that determined the realistic minimum amount of emergency allocations and demand restraints that would be politically necessary, prepared to achieve it, and then made clear that at that point the line would be drawn. No other subsidies to consumption would be tolerated. Paradoxically, planned minimum intervention may be the best way to assure maximum reliance on market forces and to avoid foolish regulations when disruptions strike.

Mitigate costs. The final task is to lessen the macroeconomic costs associated with a change in the price of an important good such as oil. The greater the oil intensity of the economy, the greater these costs; it is the level of consumption, not of imports, that matters. Unemployment and inflation greater than that dictated by the oil shock itself are inevitable because prices and wages are not flexible; little can be done about this effect. Government can, however, use monetary and fiscal policy, including redistribution of funds to those hardest hit, to offset the disruptions that otherwise would lead to avoidable unemployment.

Reducing Vulnerability: Principles for Intervention

Oil vulnerability thus has created new, though limited, roles for government. These functions fall within the mainstream of government responsibilities: they are based on traditional market-failure grounds because of externalities and the free-rider problem, on distributional concerns, and on the need for public intervention for macroeconomic stabilization. While it is not appropriate here to offer a coherent program to implement this intervention, two principles should be stressed: action must take into account the reaction of private parties and the effects of using other instruments. The key point is that policies work by affecting the price of oil and its substitutes or complements, and thus indirectly change behavior and the capital stock in sometimes unanticipated and unintended ways.

Actors within the private sector behave as they do—with regard to oil stockpiles, dual-firing capacity, natural gas reserves, and what you will—within a context that includes as an important part of the decision environment a set of existing and expected government actions. If price controls are expected, the incentive to hold inventories is reduced; if stockpile releases will lower prices during a disruption, less incentive exists to build dual-firing capacity. If the government is going to allocate coal supplies, there is every incentive not to build private stocks. If planning ahead means that someone else will get your oil, it is better to avoid the cost of preparation and let government bail you out. At the same time, some policies are complements and others substitutes. For example, the existence of stockpiles will reduce preparation for doing without oil, and thus the stockpile must be larger than the size that appears optimal if this interaction is ignored.

This dynamic interaction between private incentives and public policy, and among private actions and public policies, is at the center of effective analysis of government action. It is critical in estimating results where it counts—in terms of overall social outcomes. The process is simultaneous: one must understand and predict private behavior to design government policy, and understand how government policy will alter the way the private sector behaves to estimate whether the final outcome is desirable. Then, and only then, can reasonable conclusions be drawn about the optimal set of domestic policies to limit the avoidable costs of disruptions.

International Dimensions of Energy Vulnerability

While the goals of international action to limit oil vulnerability are reasonably clear, implementing positive programs runs head on into the international variant of the free-rider problem: sovereign governments cannot be coerced.

Goals can be grouped in three packages:

- To attack vulnerability at its source by reducing the probability, size, and duration of oil interruptions
- To achieve cooperative behavior among importers that lessens the cost of interruptions that nonetheless do occur
- To enlist oil exporters in a long-term effort to achieve stability in oil markets.

Attacking Vulnerability at Its Source

The first task in reducing vulnerability is to lessen the marginal supplier role of those countries considered most insecure. Increases in production capacity elsewhere in the world fulfill this function. They diversify supply and tend to increase the magnitude of the event necessary to trigger shortages.

As noted above, however, from a global perspective resources are wasted to the extent that more costly energy sources are discovered and produced while less costly insecure oil remains in the ground. The global gains from draining cheap resources first, if it can be done safely, provide an opportunity for beneficial cooperation among present exporters and others in the world community; they are explored below. At this point the question is how importers may limit the marginal supplier role of potentially unstable exporters at costs that are worth bearing.

From the perspective of importers as a group, any substitution for oil that can be achieved at a cost no greater than the present and (properly discounted) prospective world price of oil is worth achieving.[29] Fewer resources must be given up to produce or save a barrel domestically than would have to be traded for imports. All importers benefit through greater security and lower prices when energy supply from secure sources and conservation reflect *at a minimum* the world price. Security comes from the reduction of the marginal supplier role of those at risk and lower prices from the greater supply and lower demand for oil—the residual source of energy services. In addition, importers as a group gain in both respects when indigenous energy supply and conservation are pushed to some degree *beyond* the levels dictated by world prices. It is clear that security is enhanced, but the point is not so obvious with regard to economic welfare. The economic gains come because the price of all oil, not just the marginal barrel, is depressed by lower demand for oil exports. Because of the "bargain" that is received on the bulk of the oil imported, it is worthwhile for importers to spend more than the world price to avoid buying the marginal barrel. These gains to importers come as a transfer from exporters, who suffer an extra loss because they bear most of the extra resource cost of producing and saving energy inefficiently as well.

The benefits of restricting imports are gained by all importers, but costs are borne only by those who act. Major importers such as the United States and Japan may find it in their best interest to act unilaterally; enough benefits come their way to make bearing some costs alone worthwhile.[30] Small importers, however, are unlikely to find it in their own interest to restrict imports and, unless others participate as well, even large importers will restrict imports less than their share of the amount that would make importers as a whole better off. Thus, a cooperative restriction of imports would be in the best interest of all importers; only through firm agreements to act together can each importer be induced to restrict imports to the level that is collectively desirable. The difficulty in obtaining such agreements, unfortunately, is as clear as are the benefits that would follow. Importer organizations such as the International Energy Agency and successive summit meetings of the leaders of the major industrialized nations have called on their members to exercise restraint on imports. Success has been spotty. Beyond exhortation, no serious effort has been made at collaborative import restraint.

Another approach to limiting the major-exporter role in world oil—to foster oil and other energy production in the developing countries—is perhaps more promising. Not only would greater indigenous energy supply help relieve the developing countries' pressing oil payment burdens, but it also would help meet the security and price goals of industrialized importers. Evidence suggests, however, that oil exploration and development in the poorer countries may not even match the direct economic opportunities that exist, much less reflect the external benefits flowing to others from greater oil production. One remedy is for multinational lending institutions to facilitate oil exploration and development by providing investment funds to projects. Participation by bodies such as the World Bank also lessens political risks of oil exploitation to host governments. More needs to be known about how to make this process more effective.[31]

Vulnerability also can be reduced if ways can be found to lessen potential exogenous threats of interruption, such as those found in Middle Eastern politics and regional conflicts. Arab–Israeli hostilities twice have upset oil supplies, and energy-based economic concerns must join humanitarian and geopolitical goals as reasons to seek a settlement of this conflict. The hostilities in Lebanon only punctuate the difficulties posed for peace by the unsettled condition of the Palestinians. But the problems in the Middle East are broader and deeper. They grow out of age-old tensions and are sparked by new rivalries among nations as well, of which the Iraqi–Iranian war is only one manifestation. Religion joins politics and economic systems as a focus for disagreement and potential armed conflict in and over the oil fields. Nations also face internal stresses and even revolutions, as the case of Iran demonstrates vividly.

Oil importers in most respects are outside these processes; importing oil makes the outcomes important, but does not provide an entry point for intervention. Perhaps the best guide to policy is the watchword of classic medicine, "Above all, do no harm." There may be little importers can do to calm existing passions, but they can seek to avoid destabilizing the region further.

While supplies were tight, threats were heard that deliberate and sudden withholding of oil—"using the oil weapon"—might be used to affect U.S. policy. Such acts would be, of course, noneconomic in motivation: it is not in a seller's economic best interest to remove stock from the market or to be thought an unreliable future supplier. Though geopolitics may be the motive, the temptation to use the oil weapon still has an economic dimension. Those contemplating embargo must consider how much it costs in present and future revenues per unit of harm inflicted. This means that lowering vulnerability to disruptions through such credible instruments as strategic petroleum reserves also may lower their probability, at least as a deliberate act of exporter policy.

Finally, the oil-producing regions are another arena in which the East–West conflict may be played out. Keeping that conflict at bay is another way in which oil vulnerability can be reduced at its source. Higher economic stakes add urgency to the need to discourage Soviet intervention that would destabilize existing regimes and to stand firm against military threats on their neighbors by Soviet client states. These goals of broader U.S. foreign policy are intensified in oil-producing regions, but there—as elsewhere—the ability to control events is limited.

There are thus a number of directions U.S. policy can take that might reduce the probability, size, and duration of oil supply interruptions. In the last analysis, however, it is clear that there are no guarantees against disruptions, nor any measures that always will keep them small or short. Insecure oil is a fact of life. Vulnerability to that insecurity, however, can be limited by effective international cooperation as well as by the domestic preparations described above.

Limiting Vulnerability Through Importer Cooperation

Importers share a common burden when oil shocks occur. Output losses are followed by inflationary storms and spreading unemployment. Preparatory and remedial actions are taken within economies, but the incentives to pursue them are dampened by the fact that some of their benefits flow across national boundaries. American taxpayers understandably would be loath to build stockpiles that, when used, would mostly lower the price of oil to foreigners. No country will want to make its citizens do without oil during an interruption if that means that others get more.

Individual importers have opportunities to direct some of the costs of disruptions onto others. For example, output and employment losses can be lessened if oil imports are subsidized. Others pay part of the costs in higher oil prices. It is possible that special arrangements that give particular exporters concessions other than money can keep more oil flowing to the country willing to make such a deal, and less to others. But if importers follow a course that ignores the effect on others, all importers, including themselves, can be made worse off than they need to be. The task of achieving cooperation is not easy; it demands recognition of the gains to be had, an agreement on how they will be shared, and confidence that others will live up to agreements.

The International Energy Agency (IEA) embodies the idea of a collective approach to energy security. Its 1982 *World Energy Outlook* reminds members of the fragility of the current tranquility on the oil scene and the need to press ahead.[32] Still, the task of building stocks lags, the mechanism for avoiding a scramble for supplies in an emergency creaks, coordinated stockpile drawdown policies are not in place, and widespread concern abounds about the willingness of countries to live up to sharing agreements in a time of emergency.

As to this last point, the idea that energy security derives from the codified agreements themselves seems to mistake the underlying facts. Security rests instead on the linked community of interest of the major oil importers on a range of issues, including, at rock bottom, common defense against possible aggression. On the basis of these links it should be possible to extend the current IEA program and buttress the defense against future energy shocks. Defining the "community of interest" is not always easy or even possible, thus an energy-related issue—U.S. sanctions against Soviet energy development in response to the Polish situation—added to strains between the United States and Europe and Japan. The Soviet natural gas pipeline controversy has abated, but its memory nevertheless threatens to make further cooperation in reducing energy vulnerability more difficult.

Limiting Vulnerability Through Cooperation with Exporters

The exporter interest in stable oil markets has been disguised by the fact that oil prices have risen through sudden increases that have accompanied disruptions. In the short run, those exporters whose production was maintained clearly have gained from the disruptions. For the long run, however, uncertainty is costly for all exporters. Oil markets shrink when consumers develop substitutes and adjust consumption to account for possible disruptions. Outlets for oil, especially from the low-cost but uncertain Middle Eastern suppliers are restricted when exploration and development are directed to more secure regions. The global pie is smaller when precautionary investments must be taken. It also is smaller when uncertainty causes more costly reserves outside the insecure regions to be developed while oil that requires little labor or capital with alternative uses to produce remains in the ground. These losses in global output ultimately are shared by producers as well as consumers.

If supply shocks do occur, oil exporters also suffer from the macroeconomic losses to the world economy. Their investments abroad are put at risk and lower levels of investment dampen world prospects for economic growth in which they also share. Wide fluctuations in revenues mean that domestic investment in exporting countries may face alternating cycles of boom and stagnation, hardly conducive to good planning or efficient execution. More generally, the political instability and international scapegoating that follow from economic shocks need not leave oil exporters unscathed. These costs, borne to some degree by all exporters, are particularly onerous to countries with large reserve-to-production ratios, such as Saudi Arabia and Kuwait.

This time of relative ease in oil markets may be propitious for exploring ways exporters can join in financing and implementing energy security programs that are in their interest, too. What is required is a mechanism by which the long-term export price of oil—over which interests of importers and exporters clearly may diverge—can be separated from dampening price shocks around that trend, on which all perhaps can agree. Stockpiles of oil, built by exporters during periods of price ease and released during interruption, offer an instrument directed at stability. While use of stockpiles to dampen price spurts and drops over the long run would benefit both sides, the paradox is that each use harms one side at the expense of the other. Drawdowns during disruptions mean that exporters receive less per barrel for the oil they do sell, while in slack times rebuilding stocks raises the price for all oil consumers buy. Conflicting objectives in the short run make any agreement difficult. In addition, the record of agricultural and mineral stabilization programs suggests the temptation of parties to convert such enterprises into efforts to affect the level of prices, not just their fluctuations. When the confrontational nature of importer-exporter relations is factored in, the prospects for effective multiparty agreements seem remote.

Nonetheless, the current special circumstances make this a period when such efforts might succeed. Stockbuilding will benefit exporters by undergirding a weak market while the initial fill to get a reserve up to its optimal level will lower importer vulnerability, with oil cheap by recent standards. Producers such as Saudi Arabia or Kuwait, especially, might find producing for storage an attractive option. In addition to the gains from stabilizing prices, they (and other Gulf countries) require a minimum level of oil production to provide the associated natural gas that generates electricity and desalts water. While oil can be used to power these facilities, that means wasteful substitution of a valuable fuel for one that has few alternative uses. Both importers and producers could benefit if suitable arrangements could be reached.

The underlying direct interests of the parties suggest reason to hope that individual exporters may join importers in seeking less volatility in prices. Disruptions give exporters, too, the incentive to increase output in order to take advantage of higher prices. Individual exporters benefit when stocks acquired when oil prices are low can be released at the height of shortages. True, they lose by depressing the price of other oil they sell, but the losses are shared with other exporters. When these incentives are added to the special need of high reserve-to-production countries for stable long-term markets, a community of interest with importers can be discerned on which broader understandings and specific agreements might be based.

Conclusions

The evolution of a new relationship between energy costs and availability and the rest of the world economy is not yet complete, though nearly a decade has passed since the first oil shock. Consumption and production patterns have not fully accommodated, and the world is not wholly accustomed to the changes in wealth and influence that accompanied the manyfold increase in oil prices. The bills are not all in yet. The twin lessons of the last decade are that while initial adjustments come with surprising speed and size, the momentum of the energy-economy system is such that it cannot be turned around quickly. Long-lived capital and ingrained behavior are just too important.

Energy markets again are unsettled in 1983, but in ways remarkably different from those of not long ago. There are gluts in natural gas, oil, and coal; some analysts are predicting that the downward drift in crude oil prices will become a rout; oil company earnings have fallen back to normal levels and below; active U.S. drilling rig numbers are way down; and energy consumption—even allowing for the depressed state of the economy—is below the most optimistic expectations for conservation of a few years ago. For consumers all this is welcome news. Indeed, lower energy prices have been a significant element in reducing measured inflation, improving earnings in some industries, and perhaps in preventing a slide into depression. As the experience of the last decade suggests, however, adjustment costs will go along with these benefits: investments that looked sound with high oil prices will turn sour; oil exporters may follow oil importers as international financial casualties; and yet another bout of instability will hinder private and government planning.

Another point should be emphasized. Voices as authoritative as those who now say that energy problems are over because of the magic of the market place, a few years ago were predicting oil at $100 per barrel, and consequent disaster. Careful examination of the events of the past decade suggests that each of these views misses part of the story. In the dark days of 1979 and 1980, the $100 per barrel predictions rested on underestimations of the power of market forces in bringing new supplies on-line, in disciplining OPEC, and in inducing conservation. The pessimistic prognosticators missed the critical energy-economy tie that led to the world recession that stunted energy use and slashed OPEC oil exports. But, in turn, today's energy optimists may be mistaking a transitory, reasonably tranquil interlude and a market-driven, short-term adjustment for a long-term solution to energy problems. And, again, the importance of worldwide recession—the deepest in almost fifty years—in curtailing energy use may be underestimated. As recovery progresses, the picture can change very rapidly because of the leverage applied to OPEC oil sales.

The point is that conservation will not forever put off the day when additional energy will be required—sucking harder on the same straw will not increase supplies in perpetuity—and there is determination around the world to reverse policies that led to lower energy consumption through faltering growth. Consumption again will grow. And even though production will increase from all the energy sources that have been developed over the past few years, additional new sources will be needed in the long run.

Hindsight makes clear that the long period of expanding commercial energy use helped disguise the underlying instability of energy consumption and especially oil consumption. Oil use grew in good times and bad (though at different rates) to fuel the industrialization of economies and to replace less-convenient or less-suitable fuels. The price explosions of the 1970s removed oil's competitive edge, and the maturation of economies lowered the underlying growth trend that dominated the amplitude of cyclical swings. Now, though, the oil market will take on characteristics closer to those of the commodities addressed in the next two chapters—minerals and food. This means that oil markets will be more volatile and may increase rather than dampen the oscillations in prices and activity that accompany cyclical fluctuations. All parties—consumers and producers, importers and exporters—must adapt to a more uncertain world than the one they knew before 1973. The decade of the 1970s was not an interlude of instability now passed, but a period of transition to a permanently less-predictable energy system. A premium now has been placed on processes, institutions, and practices that can adapt to changing energy prices: a host of challenges awaits.

All of this abstracts, of course, from the central issue of oil vulnerability. It will not go away; it exposes the United States and its allies to near-intolerable economic and political risks; and private markets are not up to the task of making us as safe as we can afford to be. Steps have been taken in the right direction, but much that makes economic sense remains for government to do within the United States, and in cooperation with governments abroad, to reduce exposure to those risks we cannot avoid. Unfortunately, the urgency to act diminishes as memories of past shocks fade. A long view and dedicated statesmanship are required to create a set of energy policies and programs that will meet the diverse needs that lie ahead.

Energy did not constitute a crisis that passed into history with the 1970s. It poses, instead, a series of challenges that remain today and will be with us this decade and beyond.

Notes

1. Constructed from data (excluding centrally planned economies) presented in the World Bank, *World Development Report,* 1982 (New York, Oxford University Press, 1982) pp. 124–129.

2. As here, "OPEC" sometimes will be used as convenient shorthand for oil-exporting countries generically, but when specific comparisons are made or data are presented, the reference will be to the thirteen countries making up that organization (Algeria, Equador, Gabon, Indonesia, Iran, Iraq, Kuwait, Libya, Nigeria, Qatar, Saudi Arabia, the United Arab Emirates, and Venezuela). This subset of oil exporters is used because of consistency of data sources, not because of any conclusion that membership or nonmembership has particular behavioral or analytical significance.

3. One indication of the size of the transfer is a rough measure of the effect on oil-exporter revenues. Using 1973 annual revenues as a basis, the additional revenues of OPEC countries amounted to 1981 U.S.$1,336 billion over the 1974–81 period, an amount somewhat greater than the total GNP of France and Italy combined for 1981. [Revenue figures are found in table 2-2. GNP data are from World Bank, *1981 World Bank Atlas* (Washington, D.C., World Bank, 1982) p. 18.] Moreover, the value per barrel of the oil left in the ground rose, and more of it was left for future generations.

4. See Douglas R. Bohi and Milton Russell, *U.S. Energy Policy: Alternatives for Security* (Baltimore, Md., Johns Hopkins University Press for Resources for the Future, 1975) pp. 46–52, for a discussion of the role of implicit national discount rates in oil pricing-production decisions; and Ali D. Johany, *The Myth of the OPEC Cartel—The Role of Saudi Arabia* (New York, John Wiley, 1980) pp. 40–53, for an application of the shifting property rights hypothesis.

5. Population figures are from *1981 World Bank Atlas;* GNP data for the United States are from the Executive Office of the President, *Economic Report of the President* (February 1982) p. 234. See table 2-2 for revenue data.

6. Oil is priced in U.S. dollars. Therefore, fluctuations in the value of the dollar relative to other currencies affect what exporters get for their product in real terms and what consumers in different countries pay (see table 2-3). As to oil exporters, the mix of trading partners affects the volume of goods that dollars earned from oil can buy; those whose trade is mostly with the United States are unaffected by currency value changes.

7. Joy Dunkerley and John E. Jankowski, Jr., "The Real Price of Imported Oil," *The Energy Journal* (July 1980) pp. 113–118; and Michael J. Coda and John E. Jankowski, Jr., "The Real Price of Oil Revisited," *The Energy Journal,* October 1983.

8. Terms of trade and OPEC deflated oil prices are evident from table 2-1. The change in real revenues is presented in table 2-2. For a discussion of OPEC's purchasing power, see *Petroleum Intelligence Weekly,* October 4, 1982, p. 3.

9. Per capita data have been compiled from *World Bank Atlas* (1973) and *World Bank Atlas 1981* (1982), various pages.

10. Demand in the industrialized countries rose at 6.6 percent per year during the same period, as calculated from the American Petroleum Institute, *Basic Petroleum Data Book* vol. 1, no. 3 (1981) section VII, table 1.

11. The exception, of course, was labor. Persian Gulf countries have imported workers whose earnings have buoyed the economies of such countries as the Yemen Arab Republic and Pakistan.

12. Comparing 1979 to 1980 net oil trade balance, figures have been calculated from International Monetary Fund, *World Economic Outlook*, Occasional Paper 9 (Washington, D.C., IMF, 1982) p. 163. "Net official development assistance" is reported in the *World Development Report 1982*, p. 15. The IMF's "net oil-importing" group of countries differs slightly in composition from the World Bank's "oil-importing" group.

13. Changes in the developing countries' oil bill were calculated from the International Monetary Fund, *World Economic Outlook* (1982) p. 163. World official development assistance is reported in Organisation for Economic Co-Operation and Development, *Development Cooperation 1982 Review* (Paris, OECD, 1982) p. 180; *1980 Review*, p. 177; *1974 Review*, p. 201; and *1970 Review*, p. 171.

14. From 1975 to 1980, OPEC provided $4 to $7 billion of development assistance annually (about one-fifth of official development aid during those years), the large majority of which was provided by the Organization of Arab Petroleum Exporting Countries (*World Development Report 1982*, p. 141). Most of this aid seems to have gone to poor Arab and other Moslem countries. See, for example, OECD, *Development Cooperation* (Paris, various years).

15. Estimates of energy use (in quads) are from the U.S. Department of Energy-Energy Information Agency (DOE/EIA), *1981 Annual Report to Congress*, vol. 2 (Washington, D.C., 1982) p. 7. The U.S. real GNP is from 1982 *Economic Report of the President*, p. 234.

16. Among the demand forecasts considered: DOE-EIA, *1981 Annual Report*, vol. 3 (1982) p. 118; Exxon, *World Energy Outlook* (1975); Congressional Research Service, *Toward Project Independence* (1975); *Oil and Gas Journal* (Nov. 15, 1971) p. 124; National Petroleum Council, *U.S. Energy Outlook* (December 1972 and July 1971). Reviews of energy forecast studies can be found in Fereidun Fesharaki and T. Milo Johnson, "Short Term and Medium Term Outlook for Oil: A Review and Analysis of Recent Studies," (Honolulu, Hawaii, East-West Resource Systems Institute, January 1982); and Committee on Energy and Natural Resources, *Energy: An Uncertain Future* (Washington, D.C., December 1978).

17. The Saudi Arabian figure is derived from *World Oil*, August 15, 1982, p. 252. U.S. figures are from the U.S. Department of Energy, *Monthly Energy Review* (August 1982) p. 56; and the American Petroleum Institute, *1981 Basic Petroleum Data Book*, vol. 1, no. 3, sect. IV, tab. 6b.

18. *Petroleum Intelligence Weekly*, Jan. 10, 1982, p. 1; and *Petroleum Economist* (December 1982) p. 524.

19. In 1982 OPEC oil production was about 18 percent of world energy production. See Department of Energy–Energy Information Agency, *1981 International Energy Annual* (Washington, D.C., September 1982) p. 6, for estimates of world energy production.

20. This view is, of course, highly controversial. As time passes, however, the earlier promise of large volumes of synthetic oil from shale or coal fades and the estimates of conventional oil resources tend toward the lower end of the estimates made a few years ago. For a careful treatment of the prospects for future oil discoveries see Richard Nehering, *Giant Oil Fields and World Oil Resources* (Santa Monica, Calif., Rand Corporation, 1978); and *The Discovery of Significant Oil and Gas Fields in the United States* (Santa Monica, Calif., Rand Corporation, 1981).

21. The special problems of being "fair" in the face of the increase in energy prices are addressed by Hans H. Landsberg and Joseph M. Dukert in *High Energy Costs: Uneven, Unfair, Unavoidable?* (Baltimore, Md., Johns Hopkins University Press for Resources for the Future, 1981).

22. Excess production capacity in the United States, particularly Texas, was employed to dampen the effects of Middle East disruptions when Iranian oil supplies were restricted

in 1951, and at the time of the two Suez interruptions in 1957 and 1967. Excess capacity had disappeared by 1973, however, and incremental supplies for the world market were coming from the Middle East. For production data, see Douglas R. Bohi and Milton Russell, *Limiting Oil Imports: An Economic History and Analysis* (Baltimore, Md., Johns Hopkins University Press for Resources for the Future, 1978) pp. 22–23, and 31.

23. *Petroleum Intelligence Weekly,* Nov. 8, 1982, p. 8.

24. Among recent studies suggesting this scenario, see International Energy Agency, *World Energy Outlook* (Paris, OECD, 1982); and estimates by Helen Hughes, director of economic analysis and projections for the World Bank, "Supply and Demand for Energy: Issues for Developing Countries," paper presented at *The Oil Daily* and *The International Herald Tribune* Conference on "Oil and Money in the 80s—Energy in an Uncertain Economy," London, September 20–21, 1982.

The International Energy Agency (IEA) stated, "The current outlook for short-term stability in energy markets and the oil market in particular is deceptive because signals in today's surplus oil market do not reflect the underlying medium- and long-term trends. In fact, these trends point to recurrent oil supply stringency later in the 1980s. . ." [ibid., p. 13]. "In sum, while some structural change away from oil can be expected in the energy economy as a result of the price mechanism, the basic vulnerability of the world economy to oil supply disruptions is far from being eliminated" [ibid., p. 16]. The IEA expects world demand for oil to increase between 17 and 50 percent from 1980 to 2000, with all of the growth attributable to non-OECD countries. World oil supply, on the other hand, is expected to decline slightly or increase by only as much as 7 percent. The supply "shortfall" is expected to occur around 1990, although excess supply may dissipate by the mid-1980s. These projections imply, of course, that real oil prices will increase to balance quantities supplied and demanded.

25. This interaction is a focus of work on oil vulnerability sponsored by the Department of Energy at Resources for the Future. One product of that effort is Douglas R. Bohi's and W. David Montgomery's *Oil Prices, Energy Security, and Import Policy* (Washington, D.C., Resources for the Future, 1982).

26. See ibid., for a full discussion of policy alternatives and their implications.

27. The Reagan administration has announced that it will follow this policy with regard to releases from the Strategic Petroleum Reserve: oil would be released through auction with no more than 10 percent of monthly volume to be released directly by the Secretary of Energy through nonmarket processes. There is no formula or benchmark condition stipulated that would trigger an auction.

28. Individuals know that those who prepare will be relatively disfavored if those who do not prepare ultimately are protected from major harm. The expected value of the costs of not preparing are lower than the objective costs of not preparing because of the possibility of government help. On the other hand, the current costs of preparing are not affected by the prospect of government aid. Moreover, the marginal benefit from any future government action, *even if* applied evenhandedly, to preparers and nonpreparers alike, will be greater for those who choose to play the grasshopper rather than the ant. Consequently, the bias is toward being underprepared.

29. Consideration must be given, however, to the share of wasted resources from higher production costs that turn out to be borne by importers, rather than exporters; thus, this statement is not strictly true.

For simplicity, the discussion that follows will use world price to signify the present and prospective price at which oil flows internationally, properly discounted and adjusted for uncertainty.

30. See Bohi and Montgomery, *Oil Prices,* for an analysis of the possible role of an import fee of limited size as a component of U.S. energy and security policy with or without linkage to parallel action by other importers.

31. Research under way at Resources for the Future seeks to account for the pattern of petroleum exploration and development across non-oil developing countries and for the differing levels of activity across petroleum firms. Results will help identify efficient instruments for effecting greater investment and will clarify the rules of the private sector, host governments, home governments of petroleum firms, and multilateral international institutions. See Harry G. Broadman, "The Pattern of Petroleum Investment in Non-Oil Developing Countries: A Conceptual Framework of Company and Country Roles," in *Report of the United Nations Ad Hoc Meeting of Energy and Petroleum Experts: "Energy in the Eighties,"* (New York, United Nations, 1983).

32. International Energy Agency, *World Energy Outlook* (Paris, OECD, 1982).

chapter three

Nonfuel Minerals-The Fear of Shortages and the Search for Policies

John E. Tilton
Hans H. Landsberg

Nonfuel minerals for some years have been greatly overshadowed by energy, and with good reason. The value of their combined U.S. production, for example, does not match that of either oil or natural gas and only moderately exceeds that of coal (figure 3-1). Similarly, U.S. net expenditures on nonfuel mineral imports are only a small percentage of its net outlays for the energy it acquires from abroad. In 1980 that percentage was as low as 5 percent. In 1981 it slightly exceeded 15 percent (figure 3-2). Between 1979 and 1981 the country's annual *net* trade bill for nonfuel minerals averaged only $6 billion.

Editors' note: This chapter draws heavily on three previous works that the reader may wish to consult for additional detail: Hans H. Landsberg, "Key Elements Common to Critical Issues on Engineering Materials and Minerals," Paper presented at the Seventh Biennial Conference on National Materials Policy, July 19–22, 1982, Harpers Ferry, West Virginia; Hans H. Landsberg and John E. Tilton, with Ruth B. Haas, "Nonfuel Minerals," in Paul R. Portney, ed., *Current Issues in Natural Resource Policy* (Baltimore, Md., Johns Hopkins University Press for Resources for the Future, 1982) pp. 74–116; and John E. Tilton, *The Future of Nonfuel Minerals* (Washington, D.C., Brookings Institution, 1977).

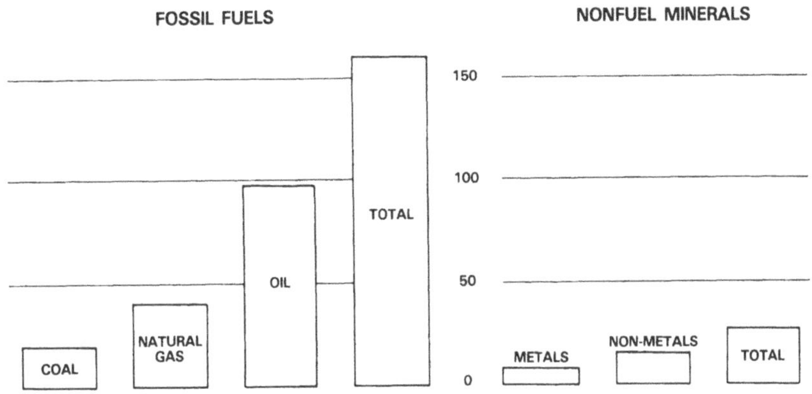

Figure 3-1. U.S. fuel and nonfuel minerals production for 1981, in billions of dollars. Data are from *Minerals Yearbook, Centennial Edition,* vol. 1, 1981, table 10; and Energy Information Administration, *Annual Report to Congress,* 1981.

Minerals differ from energy in other respects as well, and strikingly so. While in energy the spotlight shines brightest on oil, the source that is highest in value, public and policy concern in minerals focuses on a few commodities that carry a relatively low price tag. The big-ticket items—steel, copper, nickel, and aluminum—are not free of problems, to be sure, but rightly or wrongly they are not these days thought of as constituting the core of the minerals problem. Under the latter rubric fall a few metallic minerals such as cobalt, chromium, manganese, and platinum, none of which is produced in significant quantities in the United States and each of which runs up an *annual* import bill of $250 to $500 million, an amount equal to one to two *days* of imported petroleum (figure 3-3). Moreover, within the nonfuels group, the value at the mine of domestic production of nonmetallic exceeds that of metallic minerals by about two to one. Casual observers are surprised to discover that the combined value of U.S. output of stone, sand and gravel, and cement exceeds that of all metallic ores taken together (figure 3-4).

Yet another distinction is location. Secure foreign sources of minerals greatly outweigh insecure sources: Canada, Australia, Brazil, Peru, Mexico, and Venezuela are large suppliers to the United States. Canada, for example, sells to the United States more than forty, Mexico twenty, and Australia ten types of nonfuel minerals. The

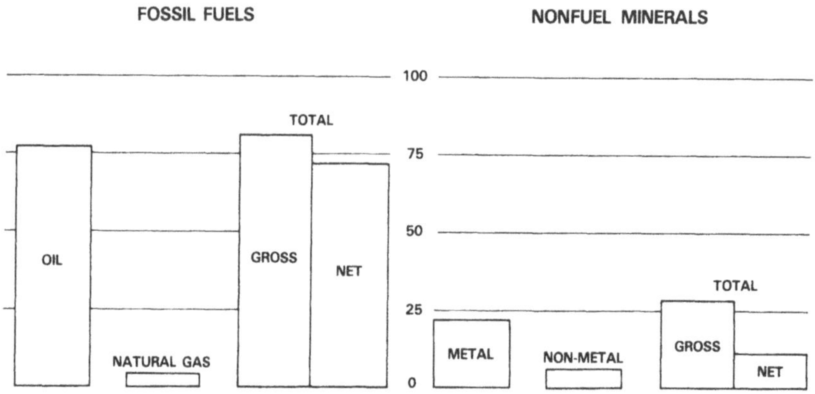

Figure 3-2. U.S. gross and net imports of fossil fuels and nonfuel minerals for 1981, in billions of dollars. Note that oil includes strategic reserve imports. Data are from *Minerals Yearbook, Centennial Edition,* vol. 1, table 10; and Energy Information Administration, *Annual Report to Congress,* 1981.

annual value of mineral imports from Canada alone represents between one-fourth and one-third of all U.S. nonfuel mineral imports. Compared with oil, then, nonfuel mineral exports to the United States are widely scattered, and to a substantial degree shipped from politically stable countries. Minerals exported from South Africa and the Soviet Union inject a note of concern, however, which will be discussed later.

Further differences are that, given the large variety of nonfuel minerals and nonmineral materials, significantly more opportunity exists for substitution among minerals than in energy. And unlike energy, nonfuel minerals are not destroyed in use. They differ in difficulty and cost of recovery, but in principle the stock is not significantly diminished, so much so that the phrase "above-ground" stocks has been coined to denote minerals in and after useful employment. Such differences caution against casual analogies between fuel and nonfuel minerals. They also suggest that conclusions and policy recommendations relevant for energy may not be appropriate for nonfuel minerals.

Despite their relatively small contribution to the gross national product and the balance of trade, a continuous flow of nonfuel minerals is essential to sustain the country's industrial base and to maintain its high standard of living. In addition, many mineral commodities produced in relatively small quantities are critical ingredients in specialty

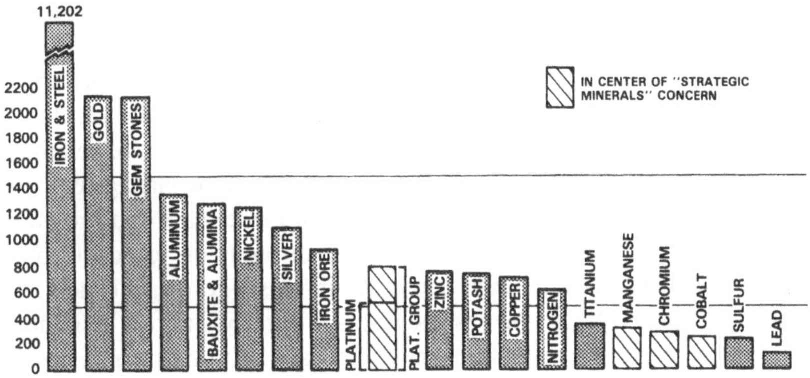

Figure 3-3. U.S. nonfuel gross mineral imports valued at more than $100 million each in 1981. Data are from *Minerals Yearbook, Centennial Edition,* vol. 1, table 10; and Energy Information Administration, *Annual Report to Congress,* 1981.

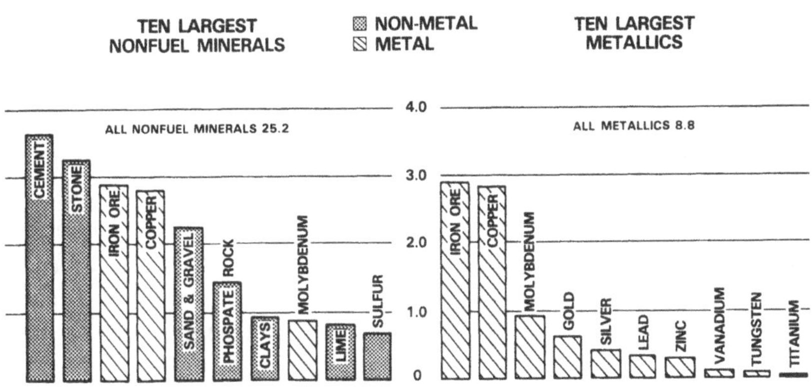

Figure 3-4. Ten largest nonfuel and metallic minerals produced in the U.S. in 1981, in billions of dollars. Data are from *Minerals Yearbook, Centennial Edition,* vol. 1, table 10; and Energy Information Administration, *Annual Report to Congress,* 1981.

materials used in the manufacture of military and other high-technology equipment.

So it is not surprising that from time to time concern arises in the United States and other industrialized countries over the future sustained availability of nonfuel minerals, even though the relatively low values suggest that the situation is amenable to management. In materials used in the manufacture of military and other high-technology equipment.

So it is not surprising that from time to time concern arises in the United States and other industrialized countries over the future sustained availability of nonfuel minerals, even though the relatively low values suggest that the situation is amenable to management. In this chapter, we first examine the nature of these fears and provide a framework for categorizing them. We then focus on efforts through mineral policies to cope with perceived threats. More specifically, we consider the evolution of past policy, the current interest in mineral self-sufficiency and protectionism, and possible alternative policies for reducing the vulnerability associated with mineral import dependence.

Shortages and Fears

To most economists, a shortage occurs when the available supply of a commodity is not sufficient at the prevailing price to satisfy demand. This means that a shortage can be eliminated simply by allowing price to rise. At some point, the higher price will adequately stimulate supply and reduce demand to bring the two into balance.

For our purposes, however, and we believe for policy decisions as well, this is too narrow a definition. While an increase in price can constrict demand to the available supply, it creates dislocation and hardship for those consumers who find they no longer can afford to buy the commodity. The higher price may bring supply and demand into balance, but at a cost that society may not be prepared to pay. So here we define shortages more broadly to include situations where demand substantially exceeds supply at the prevailing price, as well as where price rises sharply in the short run or persistently in real terms over the long run. Armed with this definition, we now proceed to examine the different reasons why mineral shortages might occur.

Resource Depletion—A Long-Run Challenge for Technology

The first step in mineral production is extraction or mining. As known deposits are depleted, new ones must be discovered or eventually shortages arise.

Two very different views of this potential problem are found in the literature. The physical view, in logic both simple and appealing, points out that since the world is finite, the available amount of a mineral commodity is a fixed stock; this is so regardless of whether we are able to see its magnitude or boundaries at any given time. Demand, on the other hand, is a flow variable that continues year after year. As a result, it is only a matter of time before the latter consumes the former and the world faces shortages. Many who adhere to this view often see the end coming sooner rather than later, as they believe demand is growing exponentially over time. Probably the best-known example of this view is found in *Limits to Growth*.[1]

The second—or economic view—of resource exhaustion does not envisage the world literally running out of mineral commodities, but instead anticipates a persistent climb in real prices over time as the better, low-cost, mines are depleted and the world is forced to turn to poorer grade, more remote, and more difficult to process deposits—all characteristics that spell rising costs.

Physical versus economic. While the ultimate effect could be just as adverse if higher prices eventually force most or all consumers to do without, there are important differences between these two perspectives. Under the economic view, shortages resulting from resource exhaustion do not occur suddenly without warning, like a speeding automobile with a broken fuel gauge that runs out of gas. Rather, they take place slowly over a long period as prices rise higher and higher. There is time to adjust to adverse consequences. More important, shortages are not inevitable: while the depletion of low-cost mineral deposits tends to push prices upward, other developments and, in particular, technological change, can partially or totally offset this tendency. And while nonfuel minerals may be widely dispersed or degraded when used, they are not destroyed, though in a given instance cost may prevent recovery.

The future availability of nonfuel minerals thus hangs largely on the outcome of the race over the long run between the cost-increasing effects of depletion and the cost-reducing effects of new technology. In the past, technology clearly has won. The real cost of producing mineral commodities has declined substantially over the last century.[2] Whether this will continue to be the case indefinitely is hotly debated. Optimists point to the past and cite earlier prophecies of doom that never have come to pass, and argue that there is no reason to believe the future will not unfold as has the past. They extol human ingenuity and note that impending shortages and higher prices encourage the technological progress needed to keep them from actually occurring. Pessimists, on the other hand, argue that the past probably is a poor guide to the future, especially in light of the tremendous growth in aggregate mineral demand during the last fifty years. They charge the optimists with blind faith in a technological "fix" to the depletion problem and consider this both simplistic and dangerous. Some also argue that the past record is deceptive in that material costs have declined largely because environmental costs associated with production have not been paid.

Which of these two views is closer to the truth, we simply do not know. Indeed, given the inherent uncertainty surrounding the future course of technological progress, there probably is no way of finding out: an early end to the debate is unlikely.

We suggest, however, that resource depletion does not pose an early threat. In particular, reserves—that is, the quantity of the mineral commodity in discovered deposits that are economic to mine given current prices and technology—are for most mineral commodities sufficient to last for the next thirty to fifty years, and for some way beyond. Moreover, where this is not the case, submarginal deposits are readily available that could be profitably exploited with only modest increases in prices.

In summary, for the near-to-middle term, future resource exhaustion does not appear to be a serious problem. Whether this happy situation will continue into the more distant future, we do not know and, indeed, cannot know. The answer depends on how successful technology is at holding the cost-increasing pressures of depletion at bay.

Investment—Two Conflicting Views

Even with high-quality reserves, shortages can arise if investment in mines and processing facilities is insufficient. Such shortages are likely to be self-correcting, for the surge in prices they generate normally will induce the needed investment. Still, new capacity cannot be built overnight. Major new mines, particularly those in remote and undeveloped areas, may take from five to seven years to bring into operation.

Ports, rail lines and roads, power plants, town sites, and other facilities may take as long to construct as the mines themselves. Naturally, if new capacity can be added by expanding existing facilities or by developing mines in established mining districts, the time required is shorter. Even in such cases, however, the delivery, installation, and breaking in of new equipment usually takes at least two years from the decision to proceed. Consequently, shortages caused by insufficient investment, though temporary and self-correcting, can cause serious problems for several years.

In examining the likelihood of such shortages, one again finds two very different views. Some observers are greatly concerned that we are not developing new mines and processing facilities to meet inescapable future demand. Many who adhere to this position are associated with major mining firms, which gives rise to suspicion that their alarm is self-serving. Yet these individuals as a rule are intimately familiar with their industries. They can cite the new mines being developed, where they are, and when they will be on-stream. They know the projects that have been stretched out, postponed, or canceled—the casualties of current market conditions. And they are familiar with how demand has grown in the past and have considerable experience in projecting it into the future. When questioned as to why they are not building more new capacity to take advantage of the favorable market conditions that the predicted shortage should produce, they point out that low profits and depressed stock prices make it impossible for them to generate internally the necessary funding. External funding from banks and other financial institutions, they argue, also is unavailable, not because the capital market has dried up, but rather because lenders—either conservative or myopic—will not lend large sums on the expectation that future requirements will drive prices up to profitable levels.

The opposing view, not surprisingly espoused by many members of the financial community, holds that the capital markets suffer from neither myopia nor other imperfections. Rather, adherents of this view maintain that external funding will be readily and amply available just as soon as there is some credible evidence that additional capacity is needed and will earn a competitive rate of return. They note that over the last decade the mineral industries in general have suffered from excess capacity, which has kept prices and profits depressed. They question whether economic growth in the industrialized nations during the rest of this century will recapture the vitality it displayed during the 1950s and 1960s. If, as is widely believed, long-run economic growth will continue to be slow, they are not convinced that investment in the mineral sector indeed is inadequate and that serious shortages loom as a result.

Both views seem plausible. However, some evidence does exist indicating that mineral firms begin more new investment projects when markets are booming and delay or cancel more projects when markets are depressed.[3] Given the long gestation period involved in bringing a new mine or processing facility on-stream, one would expect new investments to be undertaken in response solely to long-run growth trends. The fact that temporary, short-term market fluctuations appear to be important as well suggests that current conditions may unduly influence the long-run view of the market held by investors. This lends some support to the view that investors are myopic, and that by discouraging adequate investment in new mineral capacity, current depressed market conditions are sowing the seeds of future shortages.

Interruptions in Trade—The Problem of Import Dependence

Even with ample reserves and adequate investment, shortages still may occur because of interruptions in mineral production or trade, and unexpected short-term surges in demand. The first of these potential problems is examined in this section, the second in the next.

As with oil, strikes, natural disasters, embargoes, war, and a host of other factors can threaten the flow of mineral commodities to U.S. consumers. Neither domestic nor foreign supplies are immune to such interruptions. U.S. copper production, for example, in the past has been shut down for months by industry-wide strikes. Similarly, low snow cover in the Pacific Northwest on occasion has seriously curtailed hydroelectric power generation and, in turn, aluminum reduction located in that part of the country.

Still, the United States depends heavily on foreign producers for bauxite, cobalt, chromium, manganese, nickel, tin, and other important metals, as figure 3-5 indicates. Much of the recent public concern over possible shortages has focused on the possibility that these essential

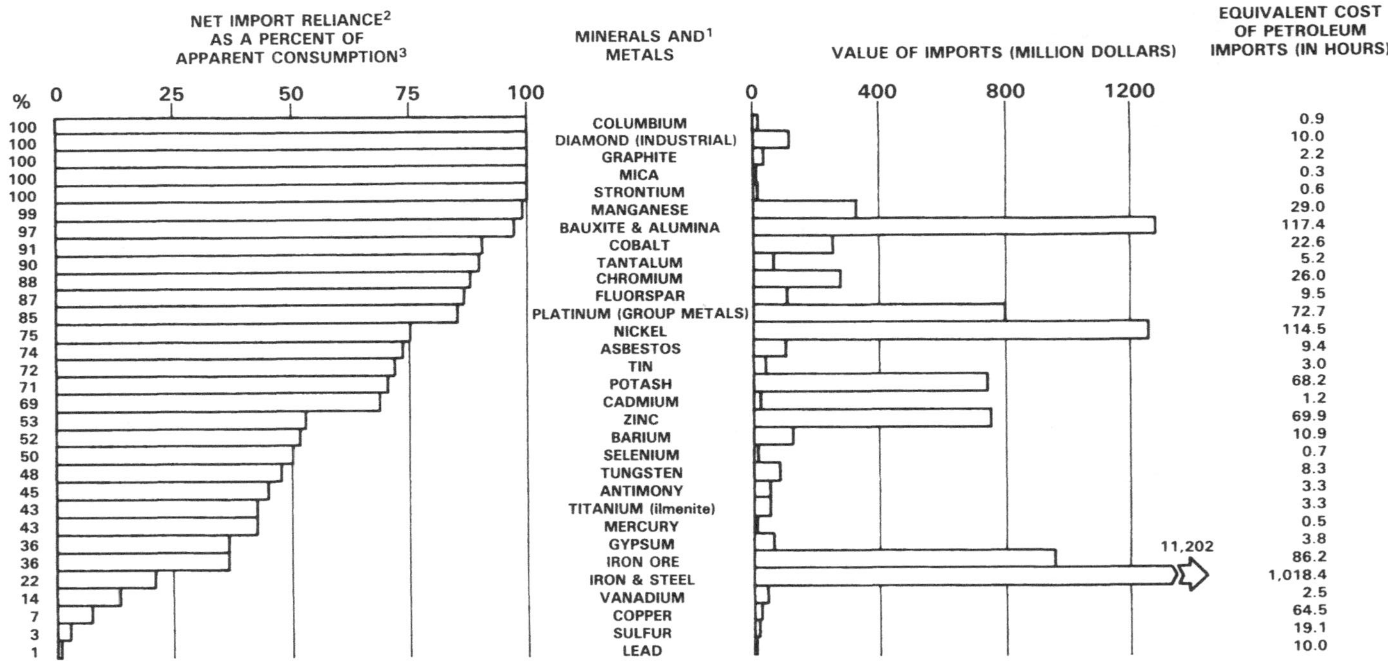

1 Substantial quantities are imported for rutile, rhenium and zircon. Data withheld to avoid disclosing company proprietary data.

2 Net import reliance = imports − exports + adjustments for government and industry stock changes.

3 Apparent consumption = U.S. primary + secondary production + net import reliance

Figure 3-5. U.S. net import reliance of selected minerals and metals as a percentage of consumption for 1981, in billions of dollars. Data for left-hand portion of figure are from U.S. Department of Interior, Bureau of Mines, *Mineral Commodity Summaries 1983*; and from John D. Morgan, "U.S. Resource Dependence," paper presented at the U.S. Department of Energy Workshop on Resources from Coal Waste and Ash, held in Reston, Virginia, June 10, 1982. Data for right-hand portion of figure are from *Minerals Yearbook, Centennial Edition,* vol. 1, table 10; and Energy Information Administration, *Annual Report to Congress,* 1981. Hourly costs of petroleum imports taken as $11 million.

imports will be restricted or cut off. Threats posed by cartels, embargoes, civil disruptions, and a "resource war" have been singled out for particular attention. Indeed, it would be fair to say that this area constitutes "the minerals problem" for many participants and observers. Because it is sensational stuff, it dominates the treatment of minerals in the news media, to the detriment, we believe, of serious long-range but more pedestrian issues, such as investment. Yet, because it looms so large in the public debate, we deal with it in some detail.

Cartels

Following the sharp increases in oil prices in 1974, it was widely believed that other mineral producers soon would join together and raise their prices. At first, a number of events seemed to bear this out. Unilateral price increases by Jamaica on bauxite and by Morocco on phosphate rock, which were followed by a number of other producing countries, appeared to confirm this concern. So, too, did the flurry of activity at about this time that led to the formation of producer associations in iron ore, bauxite, and other mineral commodities.

It now is clear that the danger of cartels in the nonfuel minerals sector was exaggerated, if not wholly unfounded. The association of major copper-exporting countries tried and failed to restrict output and maintain prices. Morocco and other phosphate rock producers soon had to rescind their price increases. Jamaica and the other Caribbean bauxite-producing countries that followed its lead in raising export taxes watched helplessly as new investment moved to other countries and their own enterprises stagnated.

To assess the future prospects for cartels, it is important to know why the expectations following the oil price increase were not realized. Are there fundamental differences between oil and the nonfuel minerals that preclude the successful creation of nonfuel cartels? Or does the explanation lie more in the long, deep slumber that overtook most mineral markets during the mid-1970s? If the latter is true, does this mean that new cartel efforts can be anticipated once the world economy recovers, and that these efforts then are likely to be more successful?

In answering such questions, we need to examine the conditions that favor the formation and maintenance of cartels. The number of such factors is quite large, but for our purposes they can be conveniently consolidated into four necessary conditions.

Limited membership. First, the number of major producing and exporting countries must be limited. The difficulties of reaching an agreement on market share, price, and other terms, acceptable to all parties, increase sharply with the number of potential participants. For most mineral commodities, this condition is satisfied, as 75 percent or more of world mine output comes from the eight largest producing countries. For chromium, nickel, platinum, tin, and many other metals, the level of concentration exceeds 90 percent.

Inelastic demand. Second, the responsiveness or elasticity of demand to changes in price must be small. Otherwise, any effort to raise price will cause a sharp loss of sales and result in lower rather than higher profits.

This condition is satisfied for most minerals in the short run—up to three years or so—but not in the long run. The reason for this is that minerals are used primarily to make metals and other materials, which in turn are inputs into final consumer and producer goods. Rarely do they constitute more than a small percentage of the total costs of the final goods in which they are embodied. The steel in an automobile, for example, accounts for no more than 5 percent to 10 percent of a car's price, and the cost of the copper or aluminum is even less significant. This means that the price of a mineral commodity can rise significantly without greatly affecting the price and hence output of the final goods from which its own demand is derived. Any significant decline in the demand for a mineral commodity caused by a rise in its price comes about through efforts by the manufacturers of final goods to substitute other, now cheaper, materials. While such substitutions often are possible, few can be made quickly. In the production of automobiles, for example, plastics and other materials can be used in place of aluminum, but such changes take several years to incorporate into the production process. It is for this reason that demand responds little to a change in price in the short run, but much more over the longer run.

Inelastic supply. Third, the responsiveness or elasticity of supply outside the control of cartel members must be small. Otherwise, an increase in price will result in a sharp loss of market share for cartel members, as producers elsewhere expand their output.

Once again, it is necessary to differentiate between the short and long term. As noted earlier, it takes from three to seven years to build new mines and processing facilities. Consequently, if an industry is operating at capacity and if available private and public stocks are small, prices can rise sharply in the short run without attracting much additional supply into the market. Over the long run, however, this is not the case, for sufficient time exists for building new capacity outside the control of the cartel.

Cartel cohesion. Fourth, the bonds among cartel members must be strong enough to prevent individual members from deserting the cartel or secretly cheating. The history of most mineral cartels again indicates that this condition may hold in the short run, but not over the long. Few cartels celebrate their fifth birthday, and only the rare exceptions have lasted ten years or longer. The DeBeers Central Selling Organization, which for almost a century now has controlled the world diamond market, truly is unique in this respect.

Just why the bonds holding most cartels together dissolve after a few years is not entirely clear. Presumably the explanation lies largely with the increased difficulties of keeping all members satisfied when the cartel's total market and profits are being eroded by the increased output of outside suppliers and by the substitution of alternative materials by consumers. And it is, of course, true that any member for which a short-term advantage is critical can reap it by violating cartel terms.

These considerations suggest that the necessary conditions for successful mineral cartels may be satisfied in the short run, except where idle production capacity or significant stocks exist outside the control of the cartel. However, since these conditions do not hold in the long run, cartels once established are likely to founder within a few years.

According to current conventional wisdom it seems irrational for producing countries to trade long-term market growth and profits for temporarily higher prices; therefore, mineral cartels are unlikely to arise even under better economic conditions. This conclusion, however, implicitly assumes that producing countries are fully aware of the adverse consequences associated with cartel participation, and that their governments are not faced with such pressing problems that they feel compelled to maximize the immediate returns from their mineral sector, regardless of the longer-term consequences. Since these assumptions may not always hold, future attempts to form mineral cartels cannot be completely ruled out.

Embargoes

Although both restrict trade, embargoes differ from cartels in several respects. Most basically, cartels are designed to raise the prices and profits realized by their members, while embargoes are imposed to relieve domestic shortages or to pressure foreign governments. For example, the United States on occasion has restricted exports of metal scrap to provide assistance to domestic consumers that claim they could not pay prices high enough to keep the scrap from flowing abroad. In a brief but celebrated instance in 1973, the U.S. government limited exports of soybeans to Japan and other countries to relieve excess pressure on supplies. This country's trading partners never fail to resurrect the memory of the soybean embargo whenever U.S. reliability as an exporter is discussed.

Examples of embargoes to pressure foreign governments are equally easy to find. For many years, exports of chromium from Rhodesia (now Zimbabwe) were proscribed under United Nations sanctions in retaliation for that country's unilateral declaration of independence from the United Kingdom and for its racial policies. In 1973 the Arab oil-exporting countries embargoed exports to the Netherlands and the United States to protest Middle East policies of these two countries and, in particular, U.S. support for Israel. The United States itself for a long time prohibited trade in minerals and other commodities with mainland China, and still imposes such sanctions on trade with Cuba.

Perhaps of greater importance than the difference in objectives is the fact that embargoes may completely cut off trade. For example, those imposed for security reasons, such as sensitive military and high-technology goods and services, are designed precisely to prevent trade in those items. Obviously, cartels will not go to this extreme, for this would eliminate rather than increase member profits.

Rarely are embargoes very successful, at least when imposed for political reasons. In addition, the long-run costs often are high for the imposing countries. Shortly after World War II, the Soviet Union embargoed manganese shipments to the United States and other Western countries. At the time, this action created considerable concern, for the Soviet Union was a major world supplier of manganese. The result, however, was not the intended change in U.S. policies, but instead the development of new manganese mines in India and elsewhere and the loss of Soviet world markets for this mineral commodity. Similarly, U.S. trade sanctions against China and Cuba appear to have had little success in achieving their desired objectives. Nor have U.S. embargoes of grain shipments to the USSR, which merely shifted Soviet purchases to alternative suppliers, as world harvests were plentiful.

Despite this poor record, embargoes continue to be attractive because they provide a visible means of expressing disapproval. When the use of stronger measures is ruled out, they give the appearance of bold action, whether or not they actually inflict any hardship on the offending country. Similarly, in countries beset by shortages and domestic inflation, they appear as constructive steps to the general public. In such situations, they even may provide some temporary relief, though at the expense of the country's foreign customers who are likely to turn to more reliable sources as soon as they are able. So, for appearances, governments are likely to continue to impose embargoes from time to time, and in the process disrupt the flow of mineral commodities in world trade.

War and Civil Disruption

Interruptions in mineral trade resulting from cartels and embargoes are the result of deliberate policies. With war and civil disruption, trade is more a hapless casualty of the breakdown in order. Twice during the late 1970s, for example, rebels based in Angola invaded the Shaba province of Zaire and overran that country's principal mining regions. As Zaire is the world's largest cobalt producer, the resulting interruption in production sent prices soaring and consumers in the United States and elsewhere scrambling for alternative sources of supply.

Past and potential flashpoints that involve mineral supply cuts could be multiplied many fold, but the point needs no belaboring. Until the world is free of war and civil disruption, production and trade of minerals—and other location-bound commodities—will continue to be vulnerable to such disturbances.

Resource War

Over the last several years, a school of thought has emerged, especially in the United States, warning of a "resource war" with the Soviet Union that threatens the security of mineral supplies from abroad, and particularly from southern Africa. In its most pointed version this is seen as part of a long-run comprehensive plan to deny the West access to the mineral wealth of southern Africa and as such is looked upon as complementing a Soviet desire to control the oil resources of the Middle East, with the invasion of Afghanistan as an alleged preliminary step.

What is not clear in these speculations is how the Soviet Union could pursue a resource war (even assuming it actually wanted to: the evidence on that point is hardly convincing, as has been pointed out in some detail in a number of recent studies).[4] Clearly the country does not have the foreign exchange to outbid western consumers for mineral supplies from southern Africa. Moreover, even if it did, this only would raise prices and stimulate expanded sources of supply in the future. Nor does it seem likely that the Soviet Union is prepared to commit large numbers of its ground forces to the region, given the logistical difficulties and the political repercussions. If it were prepared to confront the West so openly, it would find the Middle East an easier and, with its oil, a more attractive target. The ups and downs in Soviet production and trade combined with the uncertainty surrounding statistical information leave a wide-open field for speculation about motivations of policy, but the resource war interpretation seems based on thin premises.

This leaves the possibility of the Soviet Union's exploiting political and tribal conflicts and supporting indigenous opposition to South Africa and its racial policies. Such a strategy can be pursued with much less risk and cost. While offering the possibility of Soviet influence in the area, it is unlikely to lead to actual control. With fresh memories of their colonial past, the new nations of Africa will not meekly surrender their hard-won independence to a new foreign power. Moreover, their own national interests, not those of the Soviet Union or world socialism, are likely to receive highest priority. Indeed, Soviet pressures to embargo mineral exports to the West probably would backfire, for such exports correctly are perceived as a vital source of foreign exchange and economic development. In fact, when new Marxist governments have come to power—even with the help of the Soviet Union, as in Angola, Guinea, Mozambique, and Zimbabwe—they have tried to encourage, not cut, their mineral exports to and economic ties in general with the West.

For these reasons, the resource war thesis appears implausible. Not that imports of manganese, cobalt, chromium, platinum, and the other mineral commodities from southern Africa may not from time to time be interrupted. This region is going through a turbulent transition as it sheds its colonial past and moves toward black majority rule. The odds are far from trivial on wars, embargoes, rebellions, and even cartel attempts as these newly independent states grapple with their internal difficulties and hostile neighbors. But to attribute the insecurity of mineral supplies from this region to Soviet policies in pursuing a resource war misses the fundamental causes of this insecurity. This is important not for ideological reasons, but because a mistaken diagnosis is apt to lead to inappropriate and unnecessarily costly and divisive policies.

Dependence in perspective. Because it is so central to the question of import dependence, and in a way constitutes the heart of the case, we turn for a moment to figure 3-5. This impressive chart of net import reliance, published annually by the U.S. Bureau of Mines, also shows the value of U.S. imports for the same commodities for the same year. The result is illuminating. The two items showing sharp fluctuations, and of little relevance here, are gold and silver. They have been eliminated so as not to bias results in a given year. (Results will differ from year to year, but not severely.) In 1980 the total value of imports of the top five ("100 percent") entries, which contribute to the strong visual impression of dependency, was $155 million, equivalent to what in that year the United States paid for fourteen hours of oil imports. If we aggregate the import value of all minerals—omitting gold and silver—that show a dependency greater than 50 percent, the total is $6.3 billion; and of this group, four of the five largest, representing two-thirds of that total—nickel, bauxite and alumina, zinc, and potassium—present no import security problem, coming as they do from reliable sources, mostly in the Western Hemisphere. The large import values are at the bottom of the chart, but here import dependence is small and, at times, not very meaningful in the context we are dealing with. The largest item by far, of course, is iron and steel products, with a 19 percent import dependence. And here the problem lies in a different sphere altogether, that is, U.S. industrial competitiveness. What this juxtaposition suggests is that in value terms the disconcerting aspect of the graph on the left reflects only half the story. The full truth is that the country faces a number of separable supply issues involving commodities with modest dollar amounts, and thus more easily amenable to management, rather than a general threat to its survival. Modest value, while not indicating absence of a problem, does suggest manageability. While for lack of a nail a horse may be lost, it should be feasible to see to it that sufficient nails—or substitutes—are at hand. We shall return to the matter presently when we discuss remedies.

Surges in Demand

Recent concern regarding future mineral shortages has focused primarily on embargoes, civil disruptions, and other threats to the smooth flow of mineral trade, and to a lesser extent on resource depletion. Meanwhile, an equal or possibly greater problem has been largely ignored. From time to time unexpected surges in demand push demand substantially above its long-run trend. On such occasions, even though long-term supply is adequate, severe shortages occur that may persist for several years.

Military emergencies and booms in the business cycle are the most common causes of demand surges. The manufacture of arms and other military equipment requires large amounts of specialty steels, copper, aluminum, magnesium, and other metals. Consequently, when a country becomes embroiled in a war, it finds that its immediate need for mineral commodities jumps substantially above peacetime requirements.

Similarly, an economic boom can cause sharp surges in mineral demand. This is because the demand for most mineral commodities is derived largely from four end-use sectors—construction, capital equipment, automobile and other transportation, and consumer durables. The output of these sectors fluctuates with the general level of economic activity, but much more severely. When the economy is weak, their output may contract by 20 percent or more. Then when the economy recovers, their output rises sharply. However, as pointed out earlier, once full capacity is approached in the minerals industries, neither demand nor supply responds quickly to short-run changes in price. So, during an economic boom, either mineral prices must rise greatly to keep supply and demand in balance or producers must allocate or ration their available supplies. In either case, the result is shortages of mineral commodities.

Strong cyclical swings are not new to the mineral industries. For example, many mergers at the turn of the last century led ultimately to the formation of the U.S. Steel Corporation, with command at that time over some two-thirds of the country's steel-making capacity; a major force driving the mergers was the desire by producers to gain some control over the volatile steel market. Still, several recent studies suggest that the disruptions caused by such market instability may be growing more severe.

Increasing volatility. On the basis of a detailed examination of monthly fluctuations in the industrial production of the United States, Japan, Britain, France, the Federal Republic of Germany, and Italy, Chien concluded that the volatility of world industrial production, after declining from the 1950s to the 1960s, rose sharply during the 1970s.[5] His data indicate that this recent jump is due mainly to two factors—an increase in the amplitude of the business cycle of the United States and other major countries, and a growing synchronization of business cycles among the major industrialized powers. Consistent with this finding is the fact that the last major wave of mineral shortages took place during 1973 and early 1974 when the economies of the industrialized countries all were experiencing a strong economic boom. It was this situation that motivated Congress to establish the Supplies and Shortages Commission.

Some recent evidence provided by Grubb also suggests that, over the last thirty years, mineral producers, fabricators, and speculators may have reduced their willingness to vary their inventories in a countercyclical manner.[6] In particular, he found that the combined refined copper stocks held by producers and fabricators rose by an average 700 tons when monthly copper consumption dropped 1,000 tons below its five-year average in the 1950s. By the 1970s, this average inventory adjustment had declined to less than 400 tons, suggesting that when the economy is expanding the mineral stocks accumulated during the previous market downturn and available to help meet a possible surge in demand are likely to be far smaller than was the case twenty to thirty years ago. To the extent this finding holds for other countries, for other mineral commodities, and for copper at other stages of production, it raises the likelihood that future shortages due to cyclical surges in demand will be more severe and disruptive than in the past. High interest rates, by raising the cost of carrying stocks, can be a powerful factor in this area.

Summing Up

To recapitulate, there are four potential threats to the future availability of mineral commodities—resource depletion, insufficient investment in mines and processing facilities, interruptions in production and trade, and unexpected surges in demand. In designing appropriate mineral policies for coping with shortages, it is important to note that resource depletion differs from the other threats in three respects.

First, shortages caused by depletion occur over many years as real prices rise gradually, but persistently. In contrast, insufficient investment, interruptions in trade, and surges in demand are difficult to foresee and typically occur suddenly, with little or no warning. As a result, they tend to produce large jumps in price or actual physical shortages as producers ration their limited supplies.

Second, depletion-induced shortages, should they occur, can be expected to persist for many years and perhaps indefinitely, as the technology required to offset the price-increasing effects of depletion cannot be generated quickly. The other threats to the availability of minerals, while they may produce severe problems for several years, are not likely to result in lasting shortages. Insufficient mine and processing capacity can be rectified within a few years by new investment. Similarly, strikes, cartels, embargoes, civil disruptions, and other interruptions in trade usually are relatively short lived; if longer-lasting, new sources of supply can be developed in unaffected areas. While this may require the mining of submarginal deposits at somewhat higher costs, the extra costs typically are not great. Unexpected surges in demand also tend to be temporary, whether they result from booms in the business cycle or military buildups. Again, new capacity can be built within a few years if a surge in demand persists. To sum up this point: mineral shortages other than those generated by resource depletion are temporary. The disruption they cause generally is greatest immediately following their imposition, and then dissipates with time as consumers switch to alternative materials and producers develop new sources of supply.

Finally, while new technology may or may not be able to hold the cost-increasing effects of depletion at bay indefinitely, depletion is very unlikely to produce shortages over at least the next few decades, for reasons already discussed. This means that policies to prevent mineral shortages over the rest of this century, or to alleviate their adverse consequences, should be designed to cope with the short-term—though potentially severe—shortages produced by insufficient investment, interruptions in trade, and sudden demand surges.

Policy

What role might policy play in reducing the probability of shortages and their potential adverse consequences? While certain government actions can enhance security of supply with little or no cost, most such actions already have been taken. Consequently, greater security usually must be bought, in the sense that it can be acquired only at a cost. Often the cost involves consumers paying higher prices for mineral commodities.

In developing policies to protect the country from shortages, the government resembles an individual buying insurance. Two decisions are paramount. First, how much protection is desired? The answer depends on the cost of the insurance, the probability that shortages in fact will occur, the estimated adverse effects if they do, and the country's collective aversion to such risks. Second, how can the desired level of protection be acquired at least cost? Alternative policies must be reviewed in terms of the protection they afford and their costs.

Complicating the analysis is the fact that protection is a multi-dimensional concept. Some policies, for example, offer immediate—but only temporary—relief from shortages. Others have longer lead times, but once in place provide permanent protection. Certain policies may be particularly effective in preventing shortages, but of little assistance in alleviating shortages should they occur. Still other policies may be useful in dealing with particular types of shortages—those due to insufficient investment, for example—but not with other types.

How best to reconcile these various aspects or dimensions of protection, how much protection overall is desirable, and what combination of policies offers that level of protection at least cost? Judgments differ greatly, and so it is not surprising that U.S. mineral policy has been the focus of lively debate for years. The central elements of that policy since World War II, the recent push toward greater self-sufficiency, and an alternative policy that would rely primarily on stockpiling for greater protection against mineral shortages are examined next.

Past Mineral Policy

Many claim, usually with distress, that the United States never has had a mineral policy. And in the sense of a policy that is both comprehensive and internally consistent, they are right. Calls for such a policy, however, in our view are naive. As this volume documents in part, mineral policy is intertwined with energy, environmental, taxation, foreign, and defense policies. Government decisions in each of these areas affect and thus help shape mineral policy. The latter simply is not of such importance that these impacts always are consistent from the perspective of those for whom minerals form the primary focus.

In addition, the United States is among the world's largest mineral producers as well as consumers. So when mineral markets are out of balance—no matter in which direction—certain domestic groups are adversely affected and likely to turn to the political system for relief. As a result, mineral policy is not entirely shaped by fears of shortages and other consumer concerns, but rather is the product of a political process that takes into account the interests of producers and other special groups. Unsurprisingly, therefore, decisions made during shortages, when consumers are hard pressed for supplies and prices are rising sharply, often work at cross purposes with policies formulated when markets are depressed and producers in distress.

Finally, the diversity within the mineral sector calls into question the feasibility of a comprehensive policy. The governmental decisions needed for tin are quite different from those appropriate for aluminum, chromium, or molybdenum, let alone steel or copper.

For these reasons, one should not expect to find in the past—or in the future, for that matter—a comprehensive and consistent U.S. mineral policy, except perhaps as a call for a "sound and prosperous U.S. minerals industry" as is usually found as a preamble to a bill passed by Congress or a study sponsored by the White House. This does not mean, however, that there has been no policy. Myriad governmental decisions have shaped mineral policy; many are inconsistent in their effects on the mineral sector, and few are comprehensive in that they influence all or most mineral products. While an effort to identify all of these decisions would be futile, it is possible to identify two important elements or tenets that have characterized past U.S. mineral policy.

Least-cost supplies. First, the federal government not only has permitted, but even has encouraged domestic consumers to obtain mineral supplies at lowest cost, regardless of where they are found. In this connection, it has helped U.S.-based mineral firms develop mines in Canada, Peru, Australia, and other countries by offering low-cost loans, favorable taxation on foreign earnings, and insurance against expropriation and other political risks. The government also has resisted pressure from domestic producers for protection from imports. Of course, there have been some exceptions. President Eisenhower approved voluntary restrictions on lead and zinc imports in 1954, but only after efforts to appease hard-pressed domestic producers through expanded stockpile purchases proved inadequate. More recently, the carbon and specialty steel industries have received some protection but, again, only grudgingly. The more typical response to the pleas of domestic producers has not gone beyond the commissioning of studies or public pronouncements regarding the importance of a strong domestic mineral industry. The mounting of a study, which postpones at least for several years the need for any concrete action, has proved especially effective in the highly cyclical mineral sector: by the time a study is completed, markets often have recovered.

Stockpiling for security. But second, while encouraging domestic consumers to search worldwide for the lowest-cost sources for their mineral supplies, U.S. mineral policy has not completely ignored security issues. Here, however, instead of promoting self-sufficiency, it has relied primarily on stockpiling.

Legislation authorizing the federal government to stockpile mineral commodities goes back to 1939. However, wartime needs and then higher budgetary priorities during the immediate postwar period delayed the actual accumulation of stocks. It was the start of the Korean conflict in June of 1950 that eventually induced Congress to provide substantial funds for stockpiling. After this war was over, stockpiling continued, in part to bolster prices and assist the depressed domestic mineral industry. In addition, as Eckes has noted,[7] President Eisenhower took a personal interest in maintaining and increasing the stockpile:

> Through the mid-1950s, the Eisenhower administration pursued a systematic policy of building up materials stock piles, and the explanation for this vigorous policy rests with Eisenhower himself. The old general believed raw materials were better than gold; "the materials within our stock piles represent insurance against disaster," he told friends at a White House stag dinner. In 1956, when other members of the National Security Council suggested the government economize and cut back on its mobilization stock pile, because raw materials reserves caused uncertainty in the commodities markets and because these inventories seemed anachronistic in an age of atomic war, Eisenhower stood firm.

By the end of the 1950s, the United States had in place a substantial stockpile, containing many of the mineral commodities it imports.

Since that time, the stockpile has received much attention. Its size and composition, as well as the conditions under which its stocks are released, all have been subjected to considerable scrutiny and questioning. During the Kennedy and Johnson administrations, stockpiling goals were reduced by assuming a one-year, rather than a three- to five-year, war. This allowed the government to declare its holdings of some commodities in excess, and to sell off parts of the stockpile to alleviate inflation, domestic shortages, and budgetary constraints. More recently, goals again have been set on the assumption of a three-year war. As a consequence, the government now is buying certain minerals, such as cobalt, that no longer meet stockpile goals, while selling others, such as tin. At present, the composition of the stockpile is badly out of balance when measured against its objectives. Forty percent of its holdings are surplus to goal; on the other hand, its gross acquisition needs are equal to the total value of the stockpile.

Despite such imbalance, the United States has for twenty-five years possessed substantial public stocks of the major mineral commodities that it imports. These stocks over this period have constituted the country's principal defense against an unexpected interruption in foreign supplies.

Growing Pressure for Greater Self-Sufficiency

As a result of the policy just described, the United States—despite its tremendous mineral wealth—relies heavily on imports for many mineral commodities. This dependence often is contrasted with the almost complete self-sufficiency of the Soviet Union, which, for a number of reasons, has followed a deliberate policy of resource autarky. This has been quite costly in some instances, but the Soviet Union does produce almost all of the mineral products it needs.

Contrary to some popular argument, however, the United States has not passed through a major transformation over the last thirty years, moving from the status of a "have" to a "have-not" nation, from relative self-sufficiency with a net export balance to critical dependence on imports. The case for dramatically increased vulnerability—apparently perpetuated by the country's experience with petroleum—simply is not valid for the nonfuel minerals. For years, the United States has acquired most or all of its supplies of many important mineral commodities from abroad. In 1950, for example, foreign mines provided 71 percent of the country's bauxite, 77 percent of its manganese, 80 percent of its tungsten, 91 percent of its platinum, 92 percent of its cobalt, 99 percent of its nickel, and 100 percent of its chromium. This is basically the same list of commodities that generates most of the attention and concern today. Moreover, for some commodities, such as tungsten, domestic production actually has increased and import dependence decreased.

What then has changed? Why the growing interest in greater self-sufficiency?

Increasing imports. The United States has become a major importer of iron ore and steel, while in the early years after World War II it was largely self-sufficient in iron ore and a net exporter of steel. Imports of bauxite and alumina also have increased over the years, though, as noted, bauxite from abroad already accounted for nearly three-quarters of domestic consumption more than thirty years ago.

Processing shifts. Some shift in processing capacity overseas also has taken place. Aluminum now is being imported as alumina as well as bauxite. Zinc is entering the country primarily in the form of refined metal rather than as ores and concentrates, and the processing of ferroalloys increasingly is taking place abroad. Such shifts raise questions about the usefulness of domestic low-grade ores: can they bear the costs of having to be shipped abroad for processing or will they become submarginal? The transfer of processing also increases the steps in the production chain over which the United States has no direct control, and so is widely presumed to increase the country's vulnerability to supply interruptions. This, however, need not be the case. For example, alumina-producing plants are built to treat a particular type of bauxite. If imports are disrupted, other types of bauxite cannot readily be substituted. Alumina, on the other hand, is a homogenous commodity, so domestic aluminum producers can feed their smelters with alumina from anywhere in the world. This increases the choice of alternative sources of supply in the event of an interruption. The lower cost of shipping processed minerals, as they contain less waste, also tends to increase the distances from which alternative supplies can be shipped economically.

Consumption growth. Perhaps a more significant development over the last thirty years has been the substantial growth in total consumption and variety of end-use applications of a number of mineral products the United States imports. Specialty steels and other alloys, in particular, have found their way into many new high-technology and military products. Consequently, unless counteracted by stockpile releases, an interruption in supplies today could have more extensive and serious consequences than in the early postwar period.

Political changes. Finally, two important political developments apparently have increased the risks associated with mineral imports. The British Commonwealth, the French Community, and the other colonial empires that once controlled much of the developing world largely have disappeared as operating entities. The newly independent states that have emerged in their place are not always able to maintain the security and political stability required for continuous mining operations.

In addition, the economic nationalism and the fear of foreign ownership that swept through the former colonies, and also the developing countries of Latin America and developed countries, have undermined the major multinational mining corporations that for decades have dominated the production of mineral commodities, from exploration to marketing. In some countries the properties of these firms have been nationalized and now are run by the government. Other countries have forced companies to divest partial ownership to domestic interests, and still others have imposed taxes or regulations that have threatened the economic viability of company operations. For decades, the United States and other major industrialized countries have relied on these firms to supply minerals their economies require. Their waning influence calls into question their ability to continue to perform this function.

Reducing Dependence

Although import dependence is not new, the recent interest in greater self-sufficiency in the United States is not hard to understand. By increasing domestic production and lowering imports, so the argument goes, the country can reduce its exposure or vulnerability to mineral shortages.

Protectionism. Among the various measures advanced for promoting domestic production is more protection from imports. Here the advocates of greater self-sufficiency have enjoyed some success in recent years. As noted earlier, voluntary quotas, trigger prices, and other forms of protection in particular have limited specialty and carbon steel imports. There also have been efforts to curtail imports of copper, zinc, and ferroalloys, although they have not been as successful.

Tax changes. Changes in tax policy that would reduce the incentives for U.S. mineral firms to invest abroad also are being advocated.

Removal of the depletion allowance on foreign mining and the deferral of taxes on income earned abroad until repatriation have been suggested. In addition, the government-sponsored Overseas Private Investment Corporation could discontinue offering to domestic mining companies investing abroad insurance against expropriation and other political risks.

Regulatory relief. Also proposed is the relaxation of worker health and safety and environmental regulations that have been imposed on industry over the past two decades. These regulations raise domestic production costs, particularly in the mineral sector, and so increase the attractiveness of developing new mines and processing facilities abroad. The recent decision of Anaconda to close its smelter at Butte, Montana, and to ship copper concentrates to Japan for smelting often is cited as an illustration of this point, as this decision was made in part—according to the company—to avoid the costs of meeting environmental standards at the Butte smelter.

Exploiting federal lands. The opening of more federal lands to mineral exploration and development also is favored. Such lands account for about one-third of the country's surface, including Alaska, and contain many of the geologically more promising areas. Over the last twenty years, a large portion of this publicly held territory was set aside for other purposes, though there is much controversy over how large that portion really is, and how severe the restrictions are.

Direct subsidies. More direct subsidies, such as long-term purchase contracts at guaranteed prices, also are being proposed. Such supports, established in the Defense Production Act of 1950, were used in the early 1950s to encourage the rapid expansion of mineral capacity during the Korean conflict. Today they could be used to stimulate domestic production. For example, at the Blackbird cobalt deposit in Idaho, production might be feasible at four to five times the 1982 spot price of about $6 per pound of imported cobalt. Platinum in Montana is another possibility.

This incomplete list of the various measures being advocated to promote self-sufficiency illustrates the breadth and diversity of the proposed responses to the growing pressure for a major shift in U.S. mineral policy. While the push for change is largely the result of greater concern over the security of mineral supplies, it has been reinforced and accentuated by the deep and prolonged recession in mineral markets that has forced domestic producers to look to the government for relief. So far this alliance of interests has not fundamentally altered the postwar policy of relatively free mineral trade. Some government actions recently have been more consistent with the policy of self-sufficiency than free trade, but they still are exceptions. Whether the old policy eventually will crumble and make way for one emphasizing self-sufficiency is unknown. The possibility certainly exists. Whether such a change is desirable is considered next.

The Stockpiling Alternative

A shift in mineral policy toward greater self-sufficiency is attractive politically, as it would help the distressed domestic mineral sector while enhancing the security of supplies for consumers. However, greater security can be acquired only at a cost. If the country wants to buy more protection or insurance against the threat of shortages, it can do so in different ways. The rational policy is the one that buys the desired level of protection at the lowest possible price. This raises the question, Is greater self-sufficiency the most efficient way to reduce the country's vulnerability? Or can other policies provide the desired protection at lower costs?

No single answer is valid for all mineral commodities. For some, particularly those where marginal domestic producers are almost competitive with foreign firms supplying the U.S. market, self-sufficiency may be relatively inexpensive and so the best policy. However, for most mineral commodities, this is not likely to be the case.

The cost of greater self-sufficiency can be high in both economic and political terms. Poorer-grade and more difficult to process domestic ores are more expensive to exploit. The resulting higher prices that domestic consumers must pay are not merely transfer payments from them to producers, but entail real costs in the sense that more labor, capital, and energy must be consumed to meet the country's mineral needs. Nor do these costs occur just once or for a temporary period: they continue year after year for as long as public policy requires or encourages the use of poorer-quality domestic resources. The social and political costs associated with self-sufficiency involve the sacrifice of other social and public objectives, such as a cleaner environment, safer working conditions, or the preservation of wilderness areas. They also may compromise various foreign policy goals, such as the economic integration of the Western world through greater trade, or the economic development of Third World countries.

In addition, a policy of self-sufficiency implicitly equates vulnerability with import dependence. While dependence is one determinant of vulnerability,[8] there are many others—the diversity of foreign suppliers, the particular countries from which imports are obtained, domestic opportunities for material substitution, the economic and strategic importance of the end uses of mineral imports, opportunities for recycling, and the level of public and private stocks. As a result, no simple linear relationship exists between vulnerability and import dependence. Vulnerability can be high even in the absence of imports, and low despite considerable import dependence. A policy that emphasizes self-sufficiency and neglects the other factors that determine vulnerability is not likely to be efficient, but it may be very expensive.

A First Line of Defense

What then are the alternatives? In our view, stockpiling in many instances is likely to offer the most protection for the least cost, and deserves far more attention in the current debate over U.S. mineral policy than it is receiving. If the shortages the country is likely to encounter in the future will be of relatively short duration, as was concluded earlier, stockpiles are ideal. They provide an alternative source of supplies when normal channels are disrupted, without diverting investment and production from low-cost areas. Thus, they minimize the adverse effects of short-term shortages in a way that does not reduce over the longer run the efficiency of world production and the benefits of trade. Even at worst—if shortages persist for longer than expected and exhaust accumulated stockpiles—stockpiling still serves as an effective first line of defense by providing a cushion of time that allows the country to increase domestic production, to substitute alternative materials, or to develop other more permanent adjustments.

Stockpiling often is dismissed as too expensive. This conclusion, however, usually rests on a misunderstanding that confuses the real costs of stockpiling with its acquisition costs. The funds expended to purchase or accumulate stocks reflect merely the exchange of one asset for another. Just as an individual is no poorer after he draws down his savings account to buy a house, the government is no poorer after it has exchanged public revenues for copper, chromium, cobalt, or other mineral products. It may have a liquidity problem in marshaling the necessary funds, but the cost of the purchase does not reflect the consumption of real resources.

There are, of course, real costs associated with stockpiling. First, and most important, are the interest or opportunity costs on the capital tied up in maintaining stocks. Other resources also are consumed in holding and administering stockpiles. However, these costs generally are much lower than those incurred in promoting self-sufficiency, for stockpiling does not divert mining and processing from low- to high-cost areas.

This suggests that the recent push for greater self-sufficiency is misdirected, and that the postwar policy of encouraging access to low-cost mineral supplies should be continued, using stockpiles to reduce the country's vulnerability to shortages. Such a policy, however, places a heavy burden on stockpiling, and calls for a careful review of its proper role.

Disposal Policy

In the past, the government has accumulated and held strategic stockpiles under legislation that authorizes their disposal only during military emergencies. In practice, the government has used stockpile purchases and sales to stabilize mineral markets and for other non-military purposes. However, as Gauntt has shown,[9] such efforts have not been particularly successful, in part because their primary military purpose eventually constrained their flexibility in dealing with other mineral problems. A broader role for stockpiling that encompasses more than just military emergencies may be needed.

Such a change in the role of stockpiling would require a clearer specification of the purposes of stockpiling and the conditions under which sales were to occur. On occasion, the government has used stockpile sales to reduce the federal budget deficit, and to prevent producers from raising prices. Ways to prevent such unintended uses of the stockpile are needed, particularly if its role is extended.

This raises the possibility that part or all of the stocks that mineral policy supports as insurance against shortages might be held by firms and other private interests rather than by the federal government. This could make it more difficult for public officials to use the stockpile for unintended purposes, and might increase support for a broader stockpiling policy from the mineral industry, which views stockpiling with substantial misgivings. If properly designed, such a policy also might reduce the adverse effects that public stocks have on the incentives of firms to hold their stocks. The government could encourage such holdings through various measures that would help the private sector defray the additional costs.

Thus, while continuing a U.S. mineral policy that favors free access to low-cost mineral supplies worldwide and uses stockpiles to protect the country against shortages has much to commend it, stockpiling's part in the policy could be enhanced. In particular, a broader role that envisages the use of stockpiles for more than just military emergencies—as is now the case for petroleum stockpiles—with the private sector playing a more direct and active role in the actual accumulation and maintenance of stocks, seems worthy of serious consideration. More narrowly, a review is overdue of the military scenario underlying the stockpile objectives, as of the rules governing both acquisition and release decisions.

Summing Up

In summary, there are legitimate concerns for the uninterrupted flow to U.S. industry of a variety of minerals, some for the long run and some more of a short-term nature. The market provides for the needed adjustments, but temporary imbalance between supply and demand, signaled by changes in prices, cannot be avoided. When the cry of "shortages" goes up, care must be taken to diagnose the cause or causes, lest public policy, responding to pressures, thwart rather than support these adjustments.

Yet there are some potential perturbations in production or trade that call for government intervention in the form of policies that are best thought of as buying insurance. A thorough review and rethinking of concepts as well as implementation seems indicated, but all things considered, adequate stockpiles at present remain the most readily available and lowest-cost remedy.

Notes

1. Donella H. Meadows, Dennis H. Meadows, Jorgen Randers, and William H. Behrens III, *The Limits to Growth* (New York, Universe Books, 1972).

2. Harold J. Barnett and Chandler Morse, *Scarcity and Growth* (Baltimore, Md., Johns Hopkins University Press for Resources for the Future, 1963); and Harold J. Barnett, "Scarcity and Growth Revisited," in V. Kerry Smith, ed., *Scarcity and Growth Reconsidered* (Baltimore, Md., Johns Hopkins Univerity Press for Resources for the Future, 1979) pp. 163–217.

3. John E. Tilton, *The Future of Nonfuel Minerals* (Washington, D.C., Brookings Institution, 1977); and L. S. Collins, "Market Instability and Investment," *Materials and Society* vol. 5, no. 3, 1981, pp. 301–330.

4. Steven Warnecke, "International Security Implications of Soviet Energy and Raw Materials Policies," in *Probleme der Rohstoffsicherung* (Bonn, Germany, Friedrich Ebert Stiftung, 1981); William K. Severin, "Soviet Non-Fuel Minerals: Recent Trends and Prospects"; and Robert Legvold, "The Strategic Implications of the Soviet Union's Nonfuel Mineral Resource Policy," *The Journal of Resource Management and Technology* vol. 12, no. 1 (January) 1983.

5. D. Chien, "Business Cycles and Instability in Metal Markets," *Materials and Society* vol. 5, no. 3, 1981, pp. 257–265.

6. T. J. Grubb, "Metal Inventories, Speculation, and Stability in the U.S. Copper Industry," *Materials and Society* vol. 5, no. 3, 1981, pp. 267–288.

7. A. E. Eckes, Jr., *The United States and the Global Struggle for Minerals* (Austin, University of Texas, 1979).

8. Hans H. Landsberg and John E. Tilton, with Ruth B. Haas, "Nonfuel Minerals," in Paul R. Portney, ed., *Current Issues in Natural Resource Policy* (Baltimore, Md., Johns Hopkins University Press for Resources for the Future, 1982) pp. 74–116.

9. G. E. Gauntt, "Market Stabilization and the Strategic Stockpile," *Materials and Society* vol. 4, no. 2, 1980, pp. 203–209.

chapter four

The Internationalization of U.S. Agriculture

Dale E. Hathaway

The United States always has been an active exporter of agricultural products. Even well before nationhood, for example, tobacco, cotton, and other commodities were traded for manufactured goods, and as technological progress was made during the nineteenth century in transportation and refrigeration, perishables—meat, fruits, vegetables—also left American shores bound for foreign markets. By 1950, the proportion of U.S. farm income from crop and livestock exports approached 25 percent, with wheat and other grain exports playing a large and growing role.

So the United States by no means could be considered a neophyte in international agricultural trade. Still, probably no amount of experience could have prepared the country for the unprecedented and, in all likelihood, unique events of the 1970s. The resulting explosive growth in U.S. exports has changed the face of both domestic and international agriculture to the point where food policy has assumed a new and pivotal role: world events increasingly shape U.S. agriculture, and domestic economic pressures relating to food significantly influence the U.S. approach to international affairs. In "international affairs" I include foreign policy, as usually defined; international economic policy, including economic growth and exchange rates; international credit policy; and international trade policy. As each of these topics deserves a book of its own, or perhaps several, my treatment necessarily is selective. Moreover, it is not possible to show empirical relationships in many cases. I hope that the relationships outlined will interest others in exploring some of the empirical relationships more fully.

The decade of the seventies was marked by several international events, to be discussed in a few pages, that in combination precipitated dramatic changes in international agricultural trade. They also led to a significant shift in trade patterns, especially for the major trade grains—wheat and coarse grains.

Shifts in Grain Trade Patterns

In the late 1950s world trade accounted for only 2 percent of world food consumption. By the end of the 1970s, this figure had jumped sixfold—to 11 or 12 percent. The general statistic, however, conceals a growth rate that differed importantly from product to product. For example, world trade in wheat accounted for 17.8 percent of consumption in 1960–61, dropped slightly to 15.1 percent in 1970–71, and rose to 22.2 percent in 1981–82. For coarse grains (mostly corn and sorghum), the comparable figures are 5.5 percent, 7.8 percent, and 13.8 percent (in 1980–81). World trade in rice, however, has held steady at about 4 percent of consumption during the past two decades.

These figures translate into huge increases in the volume—and value—of trade. At the beginning of the 1960s, world wheat trade was 42 million tons annually. By 1970, it was averaging slightly over 50 million tons; and by 1980–81 it had nearly doubled to just under 100 million tons. World trade in coarse grains almost doubled from 24 million tons to 46 million tons during the 1960s and then leaped to 105 million tons in 1980–81. Trade volume in soybeans and soybean products also increased dramatically during the 1970s.

But it was not only the sheer volume of trade that distinguished the 1970s. Trade patterns also changed markedly over the decade as trade volume grew. Net trade in wheat and coarse grains is shown in tables 4-1 and 4-2, as are shifts in net trade positions.[1]

Table 4-1. Wheat: Net Trade Positions by Major Importer and Exporter Countries, for Selected Years
(in millions of metric tons)

Countries	*1969–72*	*1980–81*	*Change in annual net trade position over decade*
Importers			
Developed market economies			
European Community	−3.4	+9.1	+12.5
Other Western Europe	−0.3	+1.0	+1.3
South Africa	−0.1	−0.3	−0.2
Japan	−4.4	−5.6	−1.2
Centrally planned economies	−2.7	−32.9	−30.2
Eastern Europe	−3.5	−3.0	−0.4
USSR	+4.7	−15.5	−20.2
People's Republic of China	−3.9	−13.5	−9.6
Developing countries	−25.2	−41.8	−16.6
Exporters			
Developed market economies			
United States	+17.4	+41.5	+24.1
Canada	+10.5	+16.0	+5.5
Oceania	+7.4	+10.5	+3.1
European Community	−3.4	+9.1	+12.5
Argentina	+1.8	+3.7	+1.9

The developed market economies did not contribute to the growth in wheat trade. Only Japan showed an increase in annual net imports of wheat over the decade, and it was a modest 1 million tons. The Western European countries, including the European Community, switched from being net importers to net exporters of wheat as their production increased while domestic consumption remained steady.

By far the largest increase in net wheat imports was by the Soviet Union and other centrally planned economies as they moved from net imports of 2.7 million tons annually at the beginning of the 1970s to nearly 33 million tons annually a decade later. The other major growth in net imports came in the developing countries, especially the fast-growing middle-income countries and newly wealthy oil-exporting countries.

For feed grains (mostly barley and oats), the picture is somewhat different, as Japan and Western Europe substantially increased their net imports over the decade. Dominating world market growth, however, was the rapid increase in net feed grain imports by the Soviet Union, Eastern Europe, and the developing countries, including most prominently the middle-income countries.

Table 4-2. Coarse Grains: Net Trade Positions by Major Importer and Exporter Groups
(in millions of metric tons)

Countries	*1969–72*	*1980–81*	*Change in annual net trade position over decade*
Importers			
Developed market economies	−27.4	−33.5	−6.1
European Community	−13.0	−6.0	+7.0
Other Western Europe	−4.1	−8.6	−4.5
Japan	−10.3	−18.9	−8.6
Centrally Planned Economies	−3.4	−28.5	−25.1
Eastern Europe	−2.8	−10.5	−7.7
USSR	−.5	−17.0	−16.5
People's Republic of China	−.1	−1.0	−0.9
Developing Countries[a]	−4.0	−27.0	−23.0
Exporters			
United States	20.2	72.8	+50.6
Canada	3.2	3.6	+0.4
Argentina	6.8	13.9	+7.1
South Africa	2.5	4.7	+2.2
Oceania	2.3	2.5	+0.2
Thailand	1.8	2.2	+0.4

[a] Countries which were importers in both periods.

Putting it another way, the world wheat import market is now dominated by developing countries and centrally planned economies. The same two groups of countries also account for more than half of the world's coarse grain imports. Any future growth in world grain imports therefore will be determined by the economic and political considerations that influence—if not dictate—the policies and actions of the centralized, government-controlled import agencies found in all centrally planned economies and in many developing countries. Such agencies are wholly different than freely operating importing firms in market economies. What they purchase, where they purchase it, and when they purchase it are subject to internal politicoeconomic controls. It may be a matter of no little future consequence that the decision rules governing an increasing portion of the world's grain imports are not market-oriented rules in the traditional sense of the term.

The export side also has seen a dramatic shift. In the case of wheat, the United States has been by far the largest gainer among the exporters. Already the major world wheat exporter, the United States, during the 1970s, captured more than half the total *increase* in wheat trade. The emergence of the European Community as a net exporter also is significant.

But U.S. dominance of feed grain exports is even more dramatic. At the start of the 1970s, the United States exported three times as much as Argentina, its nearest competitor. By the end of the decade, U.S. exports were roughly five times those of Argentina and constituted about 70 percent of the total world feed grain exports.

The Internationalization of U.S. Agriculture

U.S. farm exports began to climb slowly in the mid-1960s as world demand generated by income and population growth exceeded domestic agricultural production. In 1960 agricultural exports were 14 percent of U.S. cash farm receipts, up from 10 percent a decade earlier, and in 1970 exports still were only 14 percent. By 1980, however, exports had more than doubled to 30 percent of total cash receipts.

The crop sector of U.S. agriculture felt the major impact. Historically—in every year from 1929 through 1973—livestock sales had substantially exceeded crop sales as a source of cash receipts to commercial farmers. Few statistics experience such sudden turnarounds: cash receipts from crops exceeded those from livestock in six of the eight years between 1974 and 1981.

At the same time, the proportion of crop income from exports was rising steadily. In 1950 it was 21 percent; in 1960, 27.5 percent; in 1970, 30.5 percent; and by 1980, the value of crop exports was 54 percent of the cash receipts from crop marketing. The comparable figures for livestock were 2 percent in 1950, 3 percent in 1960, 3 percent in 1970, and 6 percent in 1980.

Crop Response to Export Demand

The impact of export demand on individual crops is even more dramatic.

Wheat. In 1970–71 the total domestic use of wheat slightly exceeded wheat exports. In the ensuing decade, domestic wheat use increased only about 10 percent, but exports soared to more than twice domestic use. Thus, the expansion of demand for wheat in the United States was almost entirely in export markets. Exports had become the dominant factor in total demand.

Feed grain. Domestic use of feed grain for feed in 1981–82 was slightly below that of a decade earlier, but feed grain exports had grown to half of domestic feed use as compared to less than one-seventh at the beginning of the decade. Again, the total demand expansion occurred in foreign markets, and thus exports became the major determinant of prices.

Cotton. At the beginning of the 1970s cotton exports were at a low, equaling only one-third of domestic consumption. However, in 1979—for the first time in modern history—exports exceeded domestic consumption. A partial explanation lies in domestic cotton's having declined over the decade because of increasing competition from synthetics and imports.

Soybeans. Soybeans and soybean products led the export boom as the world demand for animal feed expanded. This commodity complex always has depended heavily on exports, and the degree of dependence has grown as production has expanded.

U.S. farmers were acutely aware of the changes taking place in the 1970s. They recognized that prices for their major crops were determined by events outside the United States, and they reacted—economically and politically—to these outside events.

Nor should one forget the situation that existed at the beginning of the 1970s. The demand for the major field crops had been relatively stagnant for a decade. Acreage limitation programs of one type or another had been in effect, and a substantial amount of cropland had been idled under government programs.

Suddenly, as a result of increased export demand, the domestic production-control programs were ended, and farmers responded by sharply increasing the area planted to the major export crops. Table 4-3 shows the acreage planted in the four major export crops at the beginning and end of the decade. Moreover, crop production per acre jumped a full 30 percent during the decade.

It is beyond the purpose of this chapter to evaluate the impact of this surge in foreign demand on U.S. agricultural resource use. It is difficult to believe, however, that it is possible to increase cropland harvested by some 75 million acres and to increase by about 80 million acres land planted to the four major export crops without some serious pressures on natural resources. It is especially hard to believe that

Table 4-3. U.S. Cropland Planted to Major Export Crops, 1970–82
(in millions of acres)

Crop	*1970–71*	*1981–82*
Wheat	48.7	88.9
Corn	66.9	84.2
Cotton	11.9	14.3
Soybeans	43.1	68.1
Total of four crops	170.6	255.5

wheat acreage can grow by 40 million acres without expanding into areas sensitive to wind and water erosion.

But another dramatic international event had an opposite effect. Much of the increase in U.S. agricultural production in the postwar period can be traced to sharp increases in the uses of mechanical power and agricultural chemicals. The two oil price shocks had the effect of radically revising the relative prices of petroleum-driven mechanical power and of petroleum-based agricultural chemicals. Suddenly there was a significant price incentive to switch to low-tillage cultivation and to use agricultural chemicals conservatively. It is not possible to judge with any degree of precision what effects these changes are having on natural resources and environmental quality, but they well could be important for the future.

Export Boom Effects on Price Stability and Farm Income

Throughout the 1960s agricultural producers enjoyed relatively stable prices for both agricultural inputs and outputs. Demand for farm products rose slowly, but steadily, and the index of prices received by farmers for crops never varied—up or down—by more than 4 percent from one year to the next. Of course, individual crops displayed substantially greater variation, but it was comparatively slight even in these cases. Fertilizer prices fell over the decade, those for fuel and energy were stable, and machinery prices rose slowly and steadily.

The next decade was a different story altogether. The index of prices received for crops increased by 125 percent from 1969 to 1974, fell during the following three years, rose 45 percent from 1977 until March 1981, and fell sharply in 1981 and 1982. Fertilizer prices rose 150 percent from 1970 to 1975, fell 16 percent to 1977, and then rose 44 percent after the second oil price shock. Domestic farm machinery rose threefold between 1970 and 1981.

Net farm income, which had been quite stable in the 1960s, in both nominal and real terms, climbed on a roller coaster. Nominal net farm income rose 134 percent from 1970 to 1974, and real net farm income more than doubled. Then both measures fell for four consecutive years, rose for two, and fell again in 1981 and 1982. At the beginning of the 1980s, real net farm income from farming was only one-third of the 1974 high and was the lowest recorded since World War II. Nor had the depths yet been reached: a modest recovery in 1981 was followed by a new postwar low in 1982.

Livestock producers also felt wide fluctuations in crop prices as sharp changes in feed grain and soybean prices affected the biggest cost factor in their business. Thus, most of U.S. agriculture found that much of its well-being—in terms of prices received and costs for major production inputs—was being determined by events outside the United States that to some extent were outside the control of U.S. policy.

Export Boom Effects on Balance of Payments

Agriculture historically has not been a net contributor to the U.S. balance of payments; except for the years of World War II, agricultural trade typically ran a slight deficit. The net trade balance turned positive in the 1960s, however, reached $1.5 billion in 1970, and then jumped suddenly and dramatically from 1972 to 1974. In 1980–81, exports hit $43.8 billion, and the net balance of trade for agricultural products was $26.5 billion. With the single exception of 1976–77, the net trade balance had risen every year during the decade.

This increase in export earnings from agriculture came at the exact time that the U.S. total balance-of-payments position was deteriorating because of the huge increase in world oil prices. Thus, the improving agricultural trade balance helped enormously to maintain the international economic position of the United States during the last half of the 1970s.

End of the Boom

The export boom ended, at least temporarily, with the advent of the 1980s. The 1980–81 increase in exports was one of the smallest in a decade. In 1981–82 the value of U.S. agricultural exports actually fell 8 percent from the previous year, marking the first year-to-year decline in value since 1969.

The turnaround was caused by falling prices and, in many commodities, a decline in physical volume of agricultural exports. The falloff in exports involved countries in areas that had seen the largest expansion—Eastern Europe, Latin America, Asia (including Japan), and Africa. Whereas U.S. agricultural exports to all developed countries remained unchanged, those to both centrally planned economies and to developing countries fell appreciably.

The unprecedented reversal in agricultural exports was the primary contributing factor to the significant decline in farm prices for U.S. crops, the continued slide in U.S. farm income, and a massive buildup in U.S. stocks of wheat, feed grain, cotton, rice, and soybeans. Large farm program expenditures for price supports, deficiency payments, and storage of surpluses have been, and will be, incurred. For the first time in a decade, land-diversion payments were used for 1983 crops in an attempt to reduce U.S. crop output. The decade-long policies that depended on expanding agricultural exports to absorb U.S. agricultural output have come to an end, at least for a while.

Thus, the internationalization of U.S. agriculture has been a mixed blessing and is likely to continue to be so. During many years in the decade of the seventies, the export boom brought resource returns to commercial farms to a par with the nonfarm economy. It also brought overinvestment in the crop sector, wild inflation in farmland prices, and (although there were other factors) inflated food prices. When exports slackened, sharp declines in farm prices and excess capacity in both agriculture and export-oriented agribusiness caused subtantial economic distress.

Above all, the internationalization of U.S. agriculture convinced American farmers that continued expansion of exports is crucial to their economic well-being, a conviction that is shared by many in the agribusiness community. Like their farmer customers, businesses also expanded their facilities and earnings while riding the export wave. This group—farmers and the agribusiness sector—now has a vital stake in U.S. international policies that simply did not exist before the last decade. This interest has come to bear on international issues in a major way.

Having reviewed the stunning impact of the agricultural export boom on the U.S. agricultural sector, let us now look at the international events that contributed to that explosion and to its ending—or at least slackening—in the early 1980s.

Political Events and Foreign Policy

U.S.–Soviet Relations

Détente. Establishing détente with the Soviet Union was a foreign policy milestone. This was important for food and agricultural policy because of the strong relationship in socialist countries between foreign relations and economic and trade relations: trade, credit, and investment relationships all operate within an overall context of foreign relations. Even in a market economy, private commercial transactions tend to ebb and flow with overall national relations; indeed, in some cases foreign trade, investment, and credit transactions are restricted or prohibited by law or regulation.

The beginning of détente in the early 1970s coincided with a decision by the Soviet Union to use large-scale grain imports to overcome domestic agricultural failures. It is impossible to know to what degree the two decisions were related, but in any case the United States was a major beneficiary of the grain policy change as far as exports of agricultural products were concerned. At about the same time, several of the Eastern European countries also embarked on a program of depending on grain imports as a way of increasing meat and poultry production. Soviet buying and, to a lesser extent, Eastern European imports became dominant factors in U.S. grain markets as trade relations grew between the United States and the USSR and Eastern Europe.

This new relationship was one of the fundamental factors underlying the first U.S.–Soviet grain agreement, which was signed in 1975 and covered the years 1976–81. While the agreement only obligated the Soviets to buy a minimum of 6 million metric tons of American grain (wheat and corn) every year, in practice the United States became the major supplier of the USSR as its imports soared.

The grain agreement made solid economic sense—the Soviets wanted to buy grain and the Americans wanted to sell it—but from the start the arrangement was pregnant with instability; the main question concerned the length of the gestation period. After all, these were no ordinary trading partners but the two superpowers themselves. Rival interests, fears, ideologies—the whole range of great power conflicts—threatened the agreement and eventually destroyed it for all practical purposes.

In December 1979 the Soviet Union engineered the overthrow of the government of Afghanistan, replaced it with a puppet regime, and dispatched the Soviet army to enforce the change and to quell any local resistance to it. It is not possible to divine the Kremlin's intentions or expectations regarding the takeover, but it seems safe to say that top Soviet leaders clearly did not expect to still be fighting—on a fairly major and savage scale—rebellious Afghanis four years later. And they probably did not expect the American president to react as he did, either.

The Carter embargo. On January 4, 1980, President Jimmy Carter imposed a partial embargo (not more than 8 million tons) on exports of grain to the Soviet Union as one of the sanctions to protest the invasion of Afghanistan. No action regarding agricultural trade ever has evoked so many epithets, speeches, demonstrations, and recriminations as did the Carter grain embargo.

It is not the purpose of this chapter to analyze that action and its impact on the Soviet Union; at best the economic impact was minimal, although its political effects may have been greater. Whatever the impact on the Soviet Union, however, the political fallout in the United States was immense. To say that American farmers and agribusiness groups were enraged is an understatement: the embargo was blamed for low prices, low incomes, bad weather, and sick animals. Campaigning for the presidency, Ronald Reagan attacked the embargo and promised to end it if elected.

Under the U.S.–Soviet grain agreement the Soviets agreed to buy at least 6 million tons of wheat and corn annually and could buy up to 8 million tons without consultations. At any time the United States could offer to sell, or the USSR could request to purchase, more each year. As it happened, during the first four years of the agreement the United States offered to allow purchases of more than 8 million tons every year. In each year the Soviets noted the offer and bought more in all but the first year, but they never formally agreed in writing to do so.

In the 1979–80 agreement year the Soviets had been told they could buy up to 25 million tons, and they formally had contracted for over 21 million tons when the partial embargo was imposed by President Carter. He limited export to the 8 million ton minimum under the agreement, not to the 5.5 million tons already shipped—a logic that defied U.S. allies who were asked to cooperate in the embargo. Against the counsel of most of his advisors, the president continued the embargo for the 1980–81 agreement year, even though it was obvious that no allied support for it remained and that Soviet imports were not being limited since ample supplies were available from other exporters willing to sell all they could, to the Soviets or to any other purchaser.

Ending the embargo and dealing with other issues relating to agricultural trade have not been as easy for President Reagan as candidate Reagan had suggested. The embargo was not lifted until April 1981, and then only after a long internal struggle within the new administration that pitted the Department of Agriculture and its allied commodity groups against those concerned about the continuing threat of the USSR to Poland, then (and now) in the throes of internal struggle over economic and political reform.

Once the embargo was lifted the political battle over agricultural export policy and foreign policy turned to the U.S.–USSR grain agreement which—it now seems ironically—had been concluded to stabilize grain sales. The battle involved two elements—the agreement itself and the quantity offered for sale under it. Agricultural interests lobbied for a new agreement with higher minimums, but the Reagan administration decided, for foreign policy reasons, to limit the discussions to a simple one-year extension of the existing agreement. Once this extension occurred the Soviets were offered up to 23 million tons of U.S. grain in the 1980–81 marketing year, but they chose to import only 14 million tons of that amount while their imports from other sources reached new highs. Indeed, that year the USSR imported a total of 44 million tons. The United States clearly had come to be regarded as an unreliable supplier.

Polish crisis sanctions. On December 13, 1981, the Polish government declared martial law, suspended all activities of the Solidarity trade union, arrested its leaders and other dissidents, and generally restricted internal civil liberties. These actions by the Polish authorities, and U.S. suspicion of Soviet involvement in them, led President Reagan to announce a series of sanctions including suspension of Commodity Credit Corporation credits for grain sales to Poland; suspension of negotiations on a new long-term grain agreement with the Soviet Union; and a ban on sales of all U.S. equipment for use in constructing a Soviet natural gas pipeline to Western Europe. On June 18, 1982, the president extended his ban on pipeline equipment to include sales of European technology and equipment produced under license from U.S. firms, a decision that created major foreign economic policy differences between the United States and Western Europe. Also, although the pipeline ban eventually was rescinded, it may have major long-run implications for U.S. businesses that export high technology. Again, U.S. reliability as a trading partner is perceived as shaky, with some justification.

As crop prospects improved in the United States and prices fell during 1982, the political pressures from U.S. agricultural interests to renew the grain agreement grew. Meanwhile, the Europeans—smarting over the pipeline decisions—viewed the grain issue as a major test of U.S. foreign policy consistency toward the USSR. Most U.S. exports to the Soviets, after all, are farm products.

In July 1982 President Reagan offered the Soviets a one-year extension of the current agreement and the possibility of additional sales under the extension. The White House press statement said, "The Soviets should not be afforded the additional security of a new long-term grains agreement as long as repression continues in Poland." Unstated but firmly in the minds of many farmers was a suspicion that they could not expect the security of a long-term agreement as long as exports and foreign policy are tied together by U.S. policymakers. To some, the president had the best of all worlds—a strong anti-Soviet stance and grain sales. To others, it was the worst of all worlds—a noncredible foreign trade policy that left the United States as only a residual supplier of grain to the Soviet Union.

New U.S.–USSR agreement. This situation was improved, but only marginally, by the second five-year grain agreement announced in July 1983 and signed the following month. The new agreement calls for a 50 percent boost in minimum Soviet purchases of U.S. wheat and corn: the 6 million metric ton mandatory purchase level has been upped to 9 million, and the 8 million tons allowed without further consultation has been increased to 12 million. In addition, the new agreement for the first time provides for the substitution of soybeans for grain. As much as 500,000 metric tons of soybeans or soybean meal may be purchased in place of as much as 1 million tons of grain.

Did the Reagan administration negotiate the new long-term agreement as a foreign policy signal, as a bow in the direction of economic reality, or as a political gesture to secure farm votes? Did the Soviet Union, for its part, enter into the new pact simply to assure itself of a portion of its anticipated food imports or did it intend some broader political message?

Probably each of the questions raised in the preceding paragraph can be answered in the affirmative. Obviously, complex motivations on both sides together made the conclusion of a new long-term arrangement preferable to the year-to-year sparring and feinting over annual extensions of the previous agreement. One can expect a diminution in the United States of the political posturing over food exports to the Soviet Union that has characterized recent years. The agreement certainly came as good news to American farmers and agribusinesses. And the reliability of the United States as a supplier was at least partly reestablished by the omission in the new agreement of the "escape clause" that had marked the old one. The United States no longer can reduce or suspend shipments to the USSR because of commodity shortages, a definite gain at the negotiating table for the Soviets.

Thus, both sides gained from the new agreement, which should be no surprise: mutual benefit, after all, is what trade is all about. But I should emphasize that the gains are marginal for the United States. It is true that the amounts of grain involved are larger by half than were called for in the previous agreement, but they still are small relative to what they could be in light of the fact that the United States is the world's largest seller of agricultural products and the Soviet Union its largest buyer. Twelve million metric tons, let alone 9 million, will neither make much of a dent in U.S. supplies nor place the Soviets in vulnerable dependency. And orders for significantly more than the minimum purchase levels are not likely unless and until the Soviets are satisfied that the United States will not curtail exports for foreign policy reasons. Doubtless they received U.S. assurances to that end during the negotiations leading to the new agreement, but the Soviets may be forgiven if they question how much the Americans can be believed.

Regardless of how one judges U.S. policy, the Soviet reaction is clear: the Soviets view U.S. grain exports as tied to foreign policy. Thus, despite American offers to allow much higher purchases, Soviet imports of U.S. grains during 1981–82 and 1982–83 were minimal. Moreover, the USSR has signed long-term agreements with Argentina and the Canadian Wheat Board. The new five-year agreement notwithstanding, the United States now is a residual supplier to the Soviet market and may well remain so.

Establishing Relations with China

Initiated by President Nixon's visit in 1972, in early 1979 the United States officially established diplomatic relations with the People's Republic of China. Trade expanded rapidly, U.S. agricultural exports to China grew and, in 1980, an official grain agreement was signed between the United States and China under which the Chinese agreed to purchase a minimum quantity (6 million tons) each year, and the United States agreed to supply it.

Iran and Oil

The fall of the Shah of Iran in 1979 was important not only because of its profound impact on Iran and its rippling effects throughout the Islamic world, but also because it helped create an oil market reaction that made possible the huge oil price increase of 1979–80. This second big oil shock of the decade in turn contributed to worldwide inflation, severe imbalances in the balance of payments, and the economic slowdown that began in 1980. Moreover, it shook further an already unstable region of vital importance to the United States and indeed to the entire western world.

North–South Relations

The tumultuous years of the early 1970s saw food become a contentious issue in North–South relations for the first time. Crop shortfalls in several major countries in Asia and Africa left them dependent on imports just at the time that the new world demand from the Soviets and Eastern Europe drove grain prices up. Secretary of State Kissinger considered the matter important enough to call for a UN World Food Conference in his late 1973 address to the UN General Assembly.

The World Food Conference convened in Rome in 1974. While it is difficult to trace any concrete actions to the millions of words written and spoken there, in retrospect three events of significance were moved along by the actions of governments. First, much greater emphasis was placed on agricultural development by donor nations, multilateral aid agencies, and recipient countries. Second, there was a general recognition of the need for larger and more stable food aid pledges. And a substantial push was given to the establishment of an international reserve program for wheat to provide "food security" for the world in general and the developing countries in particular.

The concern about food security helped lead to the convening in Geneva of a negotiating conference under the auspices of the UN Conference on Trade and Development. After meeting for sixty weeks during 1978 and 1979, however, the negotiating conference finally ended in failure. Its only achievement was the negotiation of a new Food Aid Convention with higher minimum levels that still were well short of the 10 million tons called for in the World Food Conference.

The food security reserve program failed because different groups of countries had different objectives for the agreement. The five major wheat exporters wanted a shared stocking system that would stabilize prices within a wide band of movement. Developing countries wanted a system that would provide food aid and balance-of-payments assistance, and that would stabilize prices within a much lower and narrower range. And the centrally planned countries wanted guaranteed supplies and relatively fixed prices.

In any event, hindsight makes clear that the proposal being negotiated would not have worked, for two reasons. First, a price-triggered reserve system would not have survived the inflation the world experienced at the end of the 1970s. And it could not function with the currency instability that has occurred since 1980.

The failure of the wheat reserve negotiations ended the major political interaction between the exporters and the developing countries just as the demand-driven boom in agricultural commodities was drawing to a close. Attention turned to economic growth, assistance levels, and debt repayment. When and if another significant price rise occurs in basic food grains, however, food security once again will become a serious North–South topic.

International Economic Events

Lagging Growth Rates

Economic growth is the engine that drives international trade, including agricultural trade, and the 1970s saw substantially less growth than the preceding years. From 1955 to 1970, gross world product (GWP) grew at a compound rate of 5.1 percent per year and the GWP per capita at 3.1 percent per year. During the 1970s, however, while GWP per capita held steady at 3.1 percent, the compound rate of growth fell to 3.8 percent.

It almost goes without saying that the declines in growth rates were not evenly distributed. It is important to emphasize the point, however, because when viewed as a total, the growth rates in developing countries did *not* decline, despite the fact that rates indeed did fall in most low-income developing countries. The major difference producing the statistical illusion was China, which experienced high rates of growth.

Growth during the 1970s in the middle-income developing countries was remarkable; as noted, this was a major factor in the expansion of world markets for agricultural products. The growth rate in the nonmarket industrial economies (the USSR and Eastern Europe) fell by more than one-half from the 1960s to the 1970s but still was respectable until the end of the decade. The industrial market economies' growth rate declined by one-third, with Japan—the leader of the industrial countries in the 1960s—producing a growth rate in the 1970s only half that of the previous decade. Even so, Japan's remained the highest sustained growth rate in the industrialized world.

Inflation

Inflation was a second major feature of the 1970s world economic scene. Annual world increases in consumer prices never had exceeded 5.1 percent from 1952 through 1969, and had averaged 4.1 percent between 1960 and 1969. Then the price escalator shifted into high gear: world consumer price increases averaged 12.1 percent for the period 1971–80. During the decade, industrial countries suffered double-digit inflation in three years and—even more painful—non-oil-exporting developing countries had average annual inflation rates exceeding 20 percent in eight years.

International Credit

As Milton Russell pointed out in chapter two, the two oil price shocks of 1974 and 1979–80 were related to and major contributors to both declining growth rates and inflation. One of the several immediate effects of soaring oil prices was a large and dramatic shift in current account trade balances. In 1970 oil-importing developing countries had a current account deficit of \$8.6 billion, \$6.9 billion of it in the middle-income group. By 1981, the comparable deficits were \$88 billion and \$75.8 billion. The industrial market economies enjoyed a current account surplus of \$12 billion in 1970; ten years later this turned to a peak deficit of \$40 billion. The industrial nonmarket economies went from modest surpluses in 1970 to deficits for most of the following years. Meanwhile, the high-income, oil-exporting countries saw a net surplus of \$2.8 billion in 1970 mushroom to an astounding near-\$115 billion in 1980.

The huge rise in the balance-of-payments deficits in developing countries, together with their desire to maintain the investment necessary for growth, led directly to staggering growth in external debt by non-oil-producing developing countries. In 1973 the long-term external debt of these countries was \$96.8 billion. By 1982, this figure had ballooned more than fivefold, to \$505 billion. Both higher borrowings and sharply higher interest rates by 1982 pushed the annual debt service payments of non-oil-producing developing countries to \$107.8 billion. The annual debt service costs for these countries now exceed 20 percent of their total export earnings.

The non-oil-exporting developing countries were not alone in embarking on a massive program of foreign borrowing. By the end of 1981, for example, the Eastern European countries had external debts estimated at \$90 billion to foreign banks and governments. The worst and most visible case was Poland, which amassed an external debt of \$26 billion; by 1981, Poland could not even service its debt. By 1982,

Rumania, the German Democratic Republic, and Hungary all were having difficulty in meeting their external debt servicing as various problems kept their export earnings low.

Some oil-exporting countries also joined the borrowing binge. By mid-1982, Mexico had a foreign debt of $80 billion and in August 1982 announced it could not meet its payments. Nigeria and some other oil-exporting countries were forced to impose stringent import controls in mid-1982 because of balance-of-payment problems arising from a shortfall in oil revenues from expected levels.

The unprecedented balance-of-payment surpluses generated by the two oil price hikes gave major impetus to the rapid expansion of debt. First, they created a huge surplus of money that had to be recycled; therefore, banks were willing and anxious to expand foreign loans. Second, many countries—developing and otherwise—saw increased borrowings as the only way to offset their outsized balance-of-payments deficits and to maintain a satisfactory internal growth rate. And it worked, at least for the short term: there is little doubt that the growth in debt helped sustain the high level of income and demand for imported food in the later years of the 1970s.

Exchange Rates

Another important international economic milestone of the 1970s was the collapse of the Bretton Woods Agreement, which had governed international exchange rates since the end of World War II. In 1973 the United States devalued the dollar, and the world entered a period of widely fluctuating exchange rates. Within those fluctuations, however, were some noticeable trends, perhaps most importantly the irregular but steady decline of the U.S. dollar from 1973 until 1980. As the International Monetary Fund noted:

> The exchange value of the U.S. dollar in 1970 (and throughout the 1960s) was very high compared with its value during the subsequent decade. The initial downward adjustment in 1971 and 1973 coincided with a period of generally less rapid inflation in the United States than in other industrial countries as a group. The real decline in the effective exchange rate for the dollar from 1970 to 1975 was thus larger than the nominal decline.
>
> The picture has been radically altered, however, by the steep rise in the U.S. dollar that began in the closing months of 1980. The real

> effective exchange rate for the U.S. dollar was slightly higher in early 1982 than in 1972, implying a considerably weaker international competitive position for U.S. manufacturers—at least as far as cost competitiveness is concerned—than they have had in any of the intervening years.[2]

Quantifying the exchange-rate effects on agricultural exports is difficult, if not impossible, because of the varying changes against individual currencies, the differences in import regulations, and so forth. It is all but certain, however, that the marked decline in the effective dollar exchange rate from 1971 through 1980 had a strong positive effect on expanding agricultural exports.

Trade Relations and International Relations

Agricultural trade policies long have departed from free-market philosophy. In general, the General Agreement on Tariffs and Trade (GATT) has found it necessary to apply different rules for agriculture than for most industrial products. As long as prices are satisfactory and exports are booming—as they were during the 1970s—international tensions arising from trade restrictions are relatively low.

The boom period allowed a satisfactory negotiation for agriculture in the Tokyo Round of trade negotiations and fostered a singular level of cooperation between the major grain-exporting nations. The relative harmony and cooperation were shattered, however, first by the U.S. grain embargo directed at the Soviet Union and then by the political pressures generated by falling exports and declining commodity prices.

When agricultural exports slacken and commodity prices fall, as they have since 1980, there is a tendency to look for ways to expand exports without regard to the finer points of international trading rules; to blame the trading practices of others for part or all of the problem of falling exports and commodity prices; and to increase import barriers in order to protect domestic producers from unsatisfactory world price trends.

Targets: The European Community and Japan

In the case of the United States, all of these tendencies have come to the fore in the last three years. Their primary focus has been the

Common Agricultural Policy (CAP) of the European Community and the import quotas maintained by Japan for a number of agricultural products, prominently citrus and beef. Constant verbal attacks by the United States on these two sets of policies have been launched from the highest levels of government. The United States has lodged five formal trade complaints with GATT and has indicated that if America loses in the General Agreement, then the GATT rules for agricultural trade must be revised or the United States will take them to be meaningless. Indeed, going beyond veiled hints that the United States will resort to export subsidies to teach the Europeans a lesson, in mid-1982 the U.S. Congress endorsed the use of export subsidies and appropriated money for their use, and a warning shot was fired across the European Community's bow in the form of a subsidized wheat flour shipment to Egypt in early 1983. In the summer of 1983 the Reagan administration played a second Egyptian card in the subsidized export to that country of 18,000 metric tons of butter and 6,000 metric tons of cheese. Egypt is a major buyer of European dairy products.

This is not the place to examine the effects of the European Community and Japanese policies, or to debate their merits. What is pertinent, however, is to note that both the CAP and the Japanese import quotas predate two multilateral trade negotiations and have survived both relatively unchanged. Thus, they are neither new nor easy to remove. Moreover, they are important political issues in the European Community and Japan. In the case of Europe, the CAP is *the* economic policy that holds the Common Market together; however costly and irrational the policy may seem to outsiders, a direct attack on its fundamental basis is a serious political issue. Coming as it does at a time of serious differences within NATO regarding European defense policy and on top of the albeit short-lived pipeline embargo, agricultural trade policy has become a major foreign policy issue between the United States and the European Community.

Similarly, citrus and beef seem insignificant in the Japanese economy, but they are far from insignificant in Japanese politics. The Japanese Diet (parliament) has a disproportionate representation from rural areas and the governing Liberal Democratic Party believes—not without justification—that if it gave in to the United States on agricultural issues it might be voted out of power. Thus, at a time when U.S.–Japanese trade relations regarding automobiles and a host of other matters are near a postwar low, the explosive agricultural trade issue is especially prickly for a government already accused of following U.S. defense and economic policy too closely.

It is puzzling that agricultural trade policy, which generally is little noted and rarely discussed at economic summit meetings, should become one of the most divisive issues between these great industrial trading partners. And it is frightening that the future of GATT—the only international institution providing rules that foster a liberal world trading system—may depend on its actions in a few agricultural issues. Ironically, these issues taken individually are not of great economic significance to the world trading system in agriculture.

The Importance of State Trading Agencies

Two other institutional developments also undermine the concept and reality of liberal trade in agriculture—the marked increase in the importance of centrally planned economies and of developing countries as importers. Both developments have sharply increased the role of state trading entities in agriculture. In centrally planned economies, of course, food imports are handled by state agencies as a matter of definition. But, as noted earlier, agricultural imports also are handled by state agencies in many developing countries. Thus, estimates are that by 1980, 81 percent of all wheat and 48 percent of all corn was imported by state agencies.

This leads to two important links in the current and prospective situation. First, state trading agencies clearly are subject to more direct political control by their respective governments than are private traders in market economies. Such state agencies simply do not respond to economic changes, either internally or externally, in the same fashion as do private firms in a market economy.

Second is the tendency of these state buying agencies to engage in bilateral supply-purchase agreements. The U.S. Department of Agriculture estimates that 40 to 50 percent of the grain now moving in world trade is covered by such agreements. By and large, the private grain trade in the United States opposes bilateral agreements, although most traders initially favored the U.S.–USSR grain agreement, because it appeared to give U.S. exporters the lion's share of the Soviet market, and despite the fact that the United States also has agreements with China and Mexico. Private traders since have learned what governments give governments for foreign policy purposes, so private trade opposition now is stronger than before to bilateral agreements which, after all, are contrary to the General Agreement on Tariffs and Trade. Oddly enough, this violation of GATT concepts is ignored by all parties, as is the role of state trading in the function of the world agriculture markets.

The United States finds itself in a sort of Catch-22 situation in international trade policies. When world commodity markets are growing and farm incomes are good, no one pays much attention to the rules of trade policy. International cooperation in good times is harmonious, and friction between agricultural trade policy and foreign policy is low. However, when international markets weaken—for whatever cause—there is a tendency to blame the trade policies of U.S. allies, and agricultural trade issues become a major foreign policy irritant. Thus, the dilemma: at the very time when international cooperation is most needed in agricultural trade it is the most difficult to achieve. Foreign policy becomes an unwilling partner of agricultural export policy driven by domestic price and income problems.

Domestic Political Reaction

This analysis suggests that the rapid internationalization of U.S. agriculture during the 1970s and the consequent substantial pressures on U.S. resources resulted from a unique combination of circumstances in international economic and political affairs. These included rapid economic growth of middle-income developing countries, fueled in part by huge increases in international borrowings; détente with the Soviet Union and its associated Eastern European countries; normalization of relations with China; the worldwide shocks generated by the two oil price hikes; and a steady decline in the real value of the U.S. dollar over most of the decade. It was inevitable that these economic changes led to changes in the political decision-making process in Washington.[3]

The Growth of Vested Interests

From their beginning in the 1930s all the way through the 1960s, the U.S. agricultural commodity programs were seen as purely domestic affairs. As such, they were largely the province of the Department of

Agriculture, and specifically the Agricultural Adjustment Administration, now the Agricultural Stabilization and Conservation Service. By the 1960s, program costs and domestic price issues brought the Bureau of the Budget, now the Office of Management and Budget, and the Council of Economic Advisors into the decision process in a major way.

The internationalization of agriculture inserted new participants into the decision process. The Department of State, the National Security Council, the Treasury Department, and the Trade Representative's Office all became heavily involved as the importance of agricultural trade to the United States grew and as it climbed higher on the list of issues among nations. As some people put it, not totally facetiously, agricultural issues have become too important to be left to the secretary of agriculture.

The increasing intrusion of parts of the U.S. government without traditional links to old-line farm organizations, commodity groups, or agribusinesses left these groups increasingly frustrated. Eventually, enough was enough and they exercised their considerable political muscle in ways that have significantly altered the flexibility of the President of the United States.

The first action affecting U.S. agricultural exports was the soybean embargo of 1973, which was undertaken for purely domestic reasons—short supply relative to demand, and inflation. Indeed, to a large extent the decision process ignored the international implications and the long-run implications for agricultural exports. The soybean embargo was short-lived—it lasted only two months—but it greatly upset both U.S. producers and major importers.

In the late summer of 1975 the United States announced a temporary embargo of grain sales to the USSR and Poland, and again short supply was given as a reason. The temporary embargo led directly to the negotiations of the first U.S.–USSR grain agreement which, as it turned out, was half of an abortive attempt at an oil–grain agreement. In any event, even though U.S. agricultural interests liked the grain agreement, with its floor under Soviet purchases, the embargo fed suspicions of farmers and the grain trade about interference with exports.

As a result, when the basic agricultural commodity legislation was renewed in the Agricultural Act of 1977, great political pressure was exerted to make short-supply embargoes all but impossible. Written into the 1977 Act were remedies so drastic and costly that almost any short-supply situation would have less effect on the economy than would the actions required by the act.

The president's ability to act for national security and foreign policy reasons was left intact in the Export Control Act of 1979. This, however, was to fall casualty to President Carter's grain embargo, as new legislation was introduced in the Agricultural Act of 1981 that substantially restricts the president's ability to impose an agricultural products embargo under the national security and foreign policy sections of the Export Control Act. This limitation on presidential powers was tacitly supported by the Reagan administration. In essence, the new law requires that drastic and extremely costly measures be taken to protect farm prices and income unless the embargo involves *all* trade or involves a country that took less than 3 percent of U.S. agricultural exports the preceding year. This was a compromise reached between farm groups that wanted all signed export contracts protected from political interference related to foreign policy issues and those that did not wish to tie the president's hands on foreign policy issues.

Whereas some argued at the time that the 1980 embargo made agricultural export policy a captive of foreign policy, many now believe the reverse—that the 1981, 1982, and 1983 Soviet grain agreement decisions made foreign policy a captive of grain export policy. Actually, neither seems to be the captive of the other. Instead, the two policies might more aptly be described as linked, each wanting to be free but bound to move in the same general direction by the realities of trade and politics.

Domestic Policy Responses

Since U.S. agricultural prices were being driven by export markets, it was possible to adopt a fundamentally new policy in the 1970s to supersede the production controls and government stockpiling that had marked the policy of the 1960s.

Initially, the only government program to supplement the market prices produced by booming exports was a target price system that guaranteed farmers a specific return per unit of commodity. It applied to export crops (except soybeans) and was viewed as a temporary measure for years when the market return was unusually low due to large crops.

After the supply-shortage concerns of the early 1970s, a second program—the farmer-owned reserve—was added in the 1977 Act. This program paid farmers to store and hold eligible crops from large harvests and to return them to the market in years of lower output and unusual foreign demand. It was anticipated that the reserve would appear and disappear and would not become a permanent fixture in the markets. Apart from a reserve of 4 million tons of wheat to be used only for emergency food aid, the government was to hold no stocks of farm commodities.

To these two programs was added a generous use of export credit, to maintain or expand foreign demand for commodities in certain countries and during periods of large supply. Essentially markets were to be allowed to clear for agricultural commodities, if not every year, at least over a two- or three-year cycle. Resource use at the farm level was to be decided by farmers, guided by market prices. Continuous, large target-price payments or a steady buildup of stocks was not anticipated. Indeed, when the Reagan administration came into office it asserted that the target-price mechanism was unnecessary and suggested it be abolished. The Congress did not see fit to follow that advice.

Obsolete Policies

It now is clear that these policies are inadequate to deal with a situation in which world output for grain and oilseeds exceeds world demand in several successive years. The large crops and weak demand of 1981 and 1982 raised the stocks held by the farmer-owned reserve to levels well above those needed to meet year-to-year fluctuations in demand or output. Target-price payments ran very high, and the government acquired large stocks of commodities from the price-support program for the first time since the 1960s. Real farm income in 1982 was the lowest in four decades, and the ratio of government subsidies to income was the highest recorded. Farm subsidies for 1983 are running at an estimated $22 billion, a level roughly equal to farm income.

All of this sums up a policy that many people—especially farmers—judge to be inadequate. It remains to be seen whether domestic policies can be changed to fit the international situation, or whether attempts will be made to influence the international situation in ways that will expand exports. Thus far, the emphasis by most farm groups has been on the latter rather than the former. There is no evidence either within government or out of it that alternatives to an export expansion policy are being considered.

Looking Ahead

What has happened to the major forces that made agricultural exports a dominant economic force during the 1970s? To the surprise of many, all of the key forces driving exports have slackened simultaneously.

Growth stops. The disruptions caused by the second oil shock and policy actions taken by authorities in the developed countries to combat the inflationary surge induced by that shock brought economic growth to a rapid halt, starting in 1980. The growth rate of the gross national product in industrial market economies fell or turned negative in 1980, 1981, and 1982. This, in turn, dampened the demand for imports of both oil and other goods. World trade, which grew steadily throughout the 1960s and 1970s, slowed appreciably in 1981, showed an absolute decline in 1982, and is undergoing only a very shaky recovery as 1983 progresses.

Credit contracts. The credit expansion that had carried the economic growth and the increasing balance-of-payment deficits of many Eastern European countries came to an end in 1981–82. As noted, the problem first became obvious regarding Poland, which by the end of 1980 had run up a foreign debt of $26 billion, partly to finance food imports. Rumania followed in 1981, and Hungary had to have special assistance in early 1982. In August 1982 Mexico stunned the financial world by announcing it could not service its then-$80 billion of external debt. Brazil's debt is comparable to Mexico's, and Argentina's—at roughly $40 billion—would appear staggering in any other company. Dozens of other countries have lesser debts, but they are only marginally less constraining. Many observers wonder whether the International Monetary Fund and the big commercial banks have sufficient resources to handle the impending crises.

All of this implies that the problem for at least the next several years is more likely to be avoiding international financial crisis and disruption than substantial expansion of credit to Eastern Europe and developing countries. It is highly unlikely that credit will be expanded to foreign borrowers in the middle-to-late 1980s at the rate that occurred in the 1970s.

Implications for agricultural exports. This will affect growth in agricultural exports in two ways. First, at least part of the real economic growth in developing countries during the past two decades has been the result of the growth of investments made by increasing external debt. A slowdown or termination of debt expansion most certainly will slow economic growth rates and, thus, the growth in import demand for farm products. It is nearly impossible to find any group—public or private—that predicts growth rates as high as those in the 1970s.

Second, the fact that centrally planned economies and most developing countries directly control agricultural imports through central buying agencies means that balance-of-payment problems are likely to lead to tighter restrictions on agricultural imports. A 1982 Department of Agriculture forecast highlights how balance-of-payment problems affect U.S. exports to financially strapped centrally planned economies:

> U.S. agricultural exports to Eastern Europe in fiscal 1981 were 23 percent below 1980. U.S. forecasts suggest that farm exports to Eastern Europe in 1982 will be about $1 billion, just over 55 percent of last year's level. . . . All countries of the region have redoubled efforts to improve their balance of trade and payments position. For the first time in recent years many of these countries appear willing to allow domestic consumption to suffer, if necessary, as commitments to improve diets through larger meat consumption are being postponed.[4]

Dollar surges. Just as international credit problems developed quickly, so did the sharp change in the value of the U.S. dollar: after nearly a decade of decline, the dollar rose sharply against other major currencies starting in late 1980. It would be foolish and almost certainly wrong to predict that the dollar will be as strong over the rest of the decade as it is in its early years, but it is hard to imagine a set of circumstances that would cause the dollar to resume the decline of the 1970s. Thus, at best, an exchange rate that encourages U.S. agricultural exports may appear only intermittently, and exchange rates may become more a source of instability than of export expansion.

The Soviet factor. If it is very difficult to foresee exchange rates, it is impossible to forecast the course of U.S.–USSR relations. One can predict, however, that the Soviet Union and the Eastern European countries will not allow themselves to become so dependent on foreign food supplies—especially on U.S. supplies—as to seriously limit their freedom of action in foreign affairs. This prediction is safe despite the second five-year grain agreement signed in August 1983. Compare, for example, the purchase levels called for in the new agreement, 9 million metric tons required and 12 million possible without consultation, with the 14 million the Soviets purchased from the United States during 1981–82 in what was considered a very disappointing year. The new five-year agreement is in the economic interest of both parties; to some extent it represents a triumph of economic reality over political ideology, and it even may be a harbinger of a second period of détente, as Carl Kaysen suggests in the next chapter. But it does not change the fact that the U.S. share of the Soviet wheat and corn market has plummeted to roughly 20 percent from the 70 percent level enjoyed before the Carter partial embargo. It is hard to imagine a situation in which the Soviets would allow themselves or their clients to become dangerously dependent on the United States.

Also, Mr. Andropov's health appears precarious, thus raising the possibility of another change in Soviet leadership in this decade. And the 1988 election will provide at least one change in U.S. leadership. In both cases the first priority almost certainly will be on defense and nuclear disarmament issues; trade and economic issues will follow rather than lead changes in relationships.

In any event, pure economic reality will prevent the Soviet Union and Eastern Europe from expanding agricultural imports in the 1980s as rapidly as they did in the 1970s. The combination of political and economic uncertainty all but guarantees limited import growth in this area of the world.

Summing Up

The events of the 1970s—so profoundly influential on U.S. agriculture—were unique and unlikely to recur in concert. As a result of rapid world economic growth, huge imports by centrally planned economies, and a decrease in the value of the U.S. dollar, American agricultural exports grew at an unprecedented rate. U.S. agriculture was transformed as foreign markets became the driving force determining farm prices and incomes.

As a natural consequence, American farmers and associated groups became acutely conscious of the importance of foreign markets to their well-being. Domestic programs were altered to be consistent with an expanding world market that would absorb U.S. increased output at favorable prices. Farmers have sought and obtained specific pieces of legislation that reduce the likelihood of interference with exports because of foreign policy and domestic inflationary decisions. They have urged and secured the expansion of credit to finance agricultural exports.

But foreign agricultural markets generally are influenced by events beyond the control of domestic agricultural interests and, indeed, often beyond the control of U.S. foreign political and economic policy. For the near- to middle-term future, at least, the United States must give more attention to developing domestic policies consistent and compatible with a highly uncertain and unstable outside world.

Notes

1. These net trade statistics are lower than gross trade figures because many countries both import and export. Net figures are a better measure of world supply and demand, and especially so of European Community supply and demand.

2. International Monetary Fund, *World Economic Outlook,* Occasional Paper 9 (Washington, D.C., IMF, April 1982).

3. These changes are outlined in more detail in Dale E. Hathaway, "Government and Agriculture Revisited: A Review of Two Decades of Change," *American Journal of Agricultural Economics* (December) 1981.

4. U.S. Department of Agriculture, Economic Research Service, *Agricultural Outlook* (Washington, D.C., USDA, August 1982) p. 10.

chapter five

U.S. Foreign Policy and Global Natural Resources

Carl Kaysen

The task of the preceding three chapters was to deal in some detail with U.S. national interests concerning energy, minerals, and food. In their examinations, the authors necessarily raised questions of foreign policy inasmuch as they are inherent in the problems at hand. Indeed, one of the major themes of this volume is that U.S. resources cannot be considered as purely domestic. The markets in which they are traded are global and, at least since the turbulent 1970s, foreign policy issues are inseparable from the demand and supply of the major natural resources.

Still, the foregoing analyses were primarily resource based. The task of this chapter is to turn the issue around and to look at it not from the resource side of the intersection of the two sets of policies but from the foreign policy side. In so doing I address three broad questions: What roles have global natural resources—in particular energy, nonfuel minerals, and food—played in the making of U.S. foreign policy? Are their roles likely to change in the near future? Should they?

Foreign Policy History

Let us begin with a capsule historical tour, and remind ourselves that before World War I the United States had little in the way of an active foreign policy in general, and no economic foreign policy as such. The nation and its government were passive and took the world as given. Americans accepted British financial and commercial dominance as more or less immutable fact. During the period up to World War I, the United States was a capital importer, both in the narrow sense of being a net capital importer on the balance-of-payments account, and in the broader sense of being an even larger capital importer through immigration.

Even in situations in which U.S. actions had large impacts on the international economy—for example, when the country became the major international wheat exporter, displacing other suppliers—Americans hardly were conscious of the changes and their international impacts, and certainly took no policy actions concerning them.

For much of the time between the two world wars, U.S. foreign economic policy (and foreign policy in general) was dominated by international financial and related trade issues that were consequences of World War I. The nation shifted from a debtor to a creditor position, and questions arising from postwar debts and reparation formed the central focus of foreign policy.

Then came the 1929 Crash and its consequences. The U.S. dollar was devalued, both financial and commercial policy were dominated totally by domestic considerations, and through its international economic policies the United States intensified the problems of the world economy rather than helping to resolve them.

The interwar period saw the beginning of the growth of U.S. business interests abroad. They were then primarily investments in the extraction of natural resources: oil in the Caribbean and Venezuela, and a little later in the Middle East; and minerals in the other Latin American countries, and to a small extent in Africa. In the late 1920s and early 1930s the concept of strategic materials became implanted in the national consciousness and, just prior to World War II, thinking about strategic materials played an important role in the U.S. attitude toward Japan. Americans saw Japanese expansion into Southeast Asia as a threat to supplies of natural rubber, on which the United States then was nearly totally dependent, as well as to tin supplies. These perceived threats certainly had an impact on the timing of the war's outbreak, although the ultimate choice was Japanese.

During the war, interest in strategic materials took on substantial importance, principally through U.S. and Allied strategic bombing campaigns and programs of preclusive purchases of ferroalloys in neutral countries aimed at restricting German access. The bombing campaigns were moderately successful, at least with respect to oil, although their success came from the damage Allied air forces inflicted on refining capacity and synthetic oil plants rather than from a direct impact on raw material supplies. The preclusive purchasing effort resulted in profits to Spanish and Turkish suppliers, and even a fall in Spanish agricultural supplies, as farmers left their farms to comb the hills for ferroalloy ores that the United States was buying at ridiculously high prices. It is not recorded that these purchases had any discernible impact on the German war effort.

After the war, the idea that some materials were strategic because of their potential significance for military production continued to concern policymakers. It certainly played a role in the origins of President Truman's Materials Policy Commission, known after its chairman, William S. Paley, as the Paley Commission. Of course, the perspective of the commission was far broader than supplies for military purposes. Rather, it framed its inquiries in terms of economic growth, and the relation of the future availability, accessibility, and price of materials to the possibilities for continuing long-term economic growth.[1]

Wartime experiences with raw material supply problems—both the direct consequences of the early Japanese advances in the Pacific and the ideas that lay behind the efforts to restrict German supplies—played a central role in the postwar U.S. decision to initiate and maintain stockpiles of strategic materials. But while the reasons for initiating the program stemmed from these experiences, the management of the stockpile as time went on responded increasingly to other pressures. Some of these arose from the concerns of U.S. domestic producers, and others from the anticipated impacts of stockpile transactions on allies and friendly nations abroad.

The Soviet Union and Security

No matter how one reckons it, the immediate postwar period ended in 1947 at the latest. From that time on, U.S. foreign policy has been dominated by security concerns, whether the time marker is taken as the Berlin blockade or as the preceding breakdown of U.S.–Soviet discussions about the German economy, when it was clear that the Soviets were interested in occupying East Germany and not in discussing the German economy as a whole. Americans first thought of their differences with the Soviet Union in global terms—as a struggle against communism. A little later, the struggle was with the Sino–Soviet bloc. Of course, the bloc never was as monolithic as it was perceived to be, and sometime in the mid-1960s America discovered that it had dissolved. For the last twenty years or so, the United States has focused more directly on the Soviet Union and less on world communism, though the Reagan administration has revived anticommunism as a major theme. Whatever their guise, these struggles and the security concerns they give rise to have been the central and dominant features of U.S. foreign policy. They have overshadowed almost all other foreign policy considerations, and they have been, in addition, the touchstone of most—though not all—foreign policy decisions.

As the leader of an alliance locked in more or less worldwide struggle, the United States maintained a continuous strand of foreign economic policy that had as its highest priority the return to and then the maintenance of a liberal international economic order. An international economy was sought in which trade and worldwide movement of capital were as free as possible, and in which firms received national treatment in other countries.

The remarkable continuity of this broad policy over the nearly forty years since the end of the war has had two foundations. The first was the security commitment. Successive U.S. administrations believed that the way to promote the strength of what was called the Free World was to argue for, and pay for, a liberal economic policy. And America did bear the costs of leadership. The second and related foundation reflected U.S. economic interests. The United States maintained the monetary standard, as Britain had in the century before. And U.S. foreign investments were growing, first in mineral and energy production, and later in manufacturing. Though still concentrated in Canada and Latin America, they spread to Europe and then to the rest of the world. Thus, U.S. security and ideological commitments have paralleled the economic interests of American business nationals.

Foreign Policy and Natural Resources

In this context, we can consider the roles played by the major natural resources, especially by those on which this volume focuses—oil and other energy resources, nonfuel minerals, and food.

Oil

We begin with oil which, in both dollar and political terms, has become the most important of all resource commodities. Until 1973, U.S. policies regarding oil were almost entirely expressions of the interests of American corporations that had concessions in Third World host countries, often jointly with British and French interests. In most situations, the Americans were latecomers in places where the British and French had led the way and played a dominant role. However, as noted earlier, this was not the case in Latin America and the Caribbean and, as time went on, the United States increasingly assumed the dominant worldwide position in terms of scale of investment.

In all these situations, the making of U.S. government policy was complicated by the fact that the primary American actors—the big international oil companies—were "private," while their British and French partners were quasi-governmental entities, which could more simply and directly call on the instruments of national policy in their dealings with host countries and each other. To further complicate the situation, the U.S. government was by no means of one mind concerning the relationship between U.S. interests and those of the major oil companies. For example, restriction of imports from the early 1950s on—made formal by executive order imposing quotas in 1959 on grounds of national security—responded mostly to domestic producer interests. It was not until 1970 that the executive branch openly questioned the national security rationale for the import quotas. The recommendation that a tariff be substituted was not accepted by Congress, and the oil crisis of 1973 made the issue moot.[2]

Oil and national security. In the period from 1945 to the early 1970s, European and Japanese consumption of oil had grown very rapidly as they substituted oil for coal in power generation and industrial use. Indeed, during the Marshall Plan period, U.S. policy helped to stimulate the substitution of oil for coal. Later, as affluence almost became the norm in the industrialized democracies, rising consumer incomes led to swelling numbers of automobiles and a corresponding growth of gasoline consumption. It was in large part because petroleum products had become so pervasively important that, after the 1973 Arab–Israeli war and the oil embargo, oil emerged as a major security interest.

When the 1973 crisis broke, the United States was in a situation in which its interests and those of its North Atlantic Treaty Organization (NATO) allies and Japan had strong elements of conflict. Americans were enormous consumers of oil—in foreign eyes, wastefully so. In particular, the Europeans and Japanese depended much more heavily on imports than did the United States. These differences were linked to and reinforced differing views on Middle East policy, and in general were a source of conflict in U.S. relations with NATO and Japan.

The differences remain. Conflicts also affect national policies with respect to other energy sources, especially nuclear power. The Europeans, with the French in the lead, saw the buildup of nuclear energy as the path to energy independence. U.S. concerns about the dangers of spreading the nuclear virus led to a more anxious view of the growth of nuclear power. The Europeans saw the American stance at best as neglectful of their interests and at worst as motivated by a desire to maintain U.S. dominance of the nuclear technology export market.

The 1979 overthrow of government in Iran and the second oil shock underlined these concerns and elevated oil and the geography of oil to an even more central position in U.S. security interests. The Middle East–Persian Gulf area now has joined the NATO central European front, Japan, Korea, and the northwest Pacific, as major foci of U.S. security interests, concern, and resources.

Securing supply. The central focus of U.S. concern with oil since 1973 has been security of supply. The U.S. economy already depends on oil imports and the dependence will grow over the longer run. Moreover, its European allies and Japan are three to five times more dependent on imported oil than is the United States and a much larger share of the imports of most of them come from the Persian Gulf area. The potential for regional wars and internal political instability to affect oil supplies has been amply demonstrated since 1973. Moreover, the threat of political or even military intervention in the area by the Soviet Union always has existed, so much so that thirty years ago the U.S. government thought that threat had turned to reality. It should not be forgotten that the Shah of Iran, who fell in 1979, took his throne in 1953 with considerable U.S. assistance, precisely because the Soviet shadow loomed large over Iran.

The Soviet Union's invasion of Afghanistan in 1979 has magnified the threat, with obvious justification. As a glance at a map of the area will attest, the Soviet Union would enjoy an enormous geographic advantage over the United States in any conventional military confrontation over the oil fields.

Because consumer outrage and economic dislocation are difficult to deal with, the threat and fear of interrupted oil supplies becomes a potent factor in international relations. Even brief interruptions of supply pose serious short-run political problems for governments. Extended supply cutoffs would precipitate correspondingly greater problems in terms of economic functioning and, for sufficiently long and large disruptions, in terms of military capacity for Western Europe and Japan. Even for the United States, the direct effects of a prolonged cutoff of Persian Gulf supplies—say, a year or more—would be considerable; the indirect security effects through U.S. allies might well be profound. These eventualities, although unlikely to come to pass, are dark political clouds that hang over and influence U.S. security policy, both in terms of military plans and deployments and of diplomacy.

As of mid-1983, the cumulative effect of three years of recession-induced decline in demand for oil on its price and on the cohesion of the Organization of Petroleum Exporting Countries (OPEC) may have pushed back these fears from the center of interest of policymakers in the United States and elsewhere. But this respite, if it exists, is sure to be temporary.

The petrodollar problem. The oil price revolution of the 1970s had a further range of impacts on U.S. foreign policy, less salient and central than those in the security area; these arose from its economic consequences, as was discussed by Milton Russell in chapter two. At first, there was much discussion of the "recycling" problem: how could the international financial system handle the large accumulations of liquid assets in the hands (or foreign balances) of the newly rich oil exporters, especially the thinly populated countries of the Persian Gulf? Some feared the breakdown of the whole international payments mechanism. But the anticipated difficulties never really materialized. Until recently, the international banking system showed itself equal to the smooth absorption and transfer of these vast additions to its short-term accounts.

Ironically, the current anxiety springs from the opposite direction. The fall in the earnings of oil exporters has put in doubt the capacity of several of them, especially Mexico, Nigeria, and Venezuela, to service their debts to American and European banks. International measures of response, including cooperative action by the bankers and special help from the International Monetary Fund (IMF) have been put in motion, and it is to be expected that the anticipated worst will not occur.

Poorest of the poor. Another, and more important, effect of the change in the world oil market has been the further impoverishment of the great majority of Third World countries that nature has not endowed with huge amounts of oil. Since concern for their economic health plays a relatively small role in U.S. foreign policy—rightly or wrongly—this change has not had a major effect on that policy. What could be done? Many would argue for substantial, if not corresponding, increases in U.S. economic assistance, either in bilateral form or in contributions to multilateral aid-giving agencies. In any event, however, it is not possible for the United States alone to eradicate global poverty.

Finally, as documented in earlier chapters, the explosion of oil prices over the last decade reverberated worldwide in the surge of inflation and the slowdown of economic growth. Just how important this contribution has been in quantitative terms is a matter of some dispute among economists, perhaps less so in respect to inflation and more in respect to slow growth. Also a matter of dispute is how much blame to attribute to the price changes themselves, viewed as exogenous events outside the control of the importing countries, and how much to the policy responses—and nonresponses—in those countries. But again, regardless of how the blame is sorted out, the poorest countries have been hit very hard.

Uranium and Coal

Oil clearly is the primary global energy resource, but two other fuel minerals deserve at least brief mention.

Uranium. As the 1960s gave way to the 1970s, U.S. fears arose about the long-term adequacy of domestic and international uranium supplies for the then-predicted growth of demand from nuclear power plants. Dating back to still an earlier period, the United States promoted exports of nuclear power plant equipment through subsidies and commitments to provide fuel (enriched uranium) for U.S.-built reactors. Aimed at U.S. allies and "friendly" neutrals, the rationale for the program was a mixture of propaganda ("Atoms for Peace"), economic aid, and promotion of the domestic industry.

A variety of concerns has clouded the future of nuclear power generation, including slower growth in the demand for power, anxieties about safety and waste disposal, broader and more diffuse antinuclear sentiment, and at times worries about possible links between power generation and weapons production. Taken together with diminution of the fears for the adequacy of uranium supplies, these have greatly diminished the foreign policy significance of uranium.

Coal. Coal has received some attention as a potential substitute for oil on both the national and international scenes. But coal may never realize its potential; indeed, it already may have peaked. Although ample global reserves exist, production in fact has been static or declining in many producing countries. Large changes in consumers' habits, substantial investments, and either technological or policy changes that made the serious environmental consequences of significant increases in coal consumption acceptable all would have to occur before substitution of coal for oil on a large scale is likely.

Nonfuel minerals

The nonfuel minerals in recent years have not had anything like oil's depth and breadth of impact on U.S. foreign policy thinking. At best they have been an episodic minor theme, which has appeared in two variations. The first is the problem of protecting U.S. investors abroad against nationalization, with the Chilean and Peruvian copper mines as leading examples.

The other variation is that of fluctuations in international markets. As John Tilton and Hans Landsberg indicate in chapter three, the markets for minerals tend to very wide price fluctuations that greatly affect the prosperity of American domestic producers and the fortunes of some friendly countries that are major exporters. These difficulties lead to pressures—especially from Third World producers—for U.S. participation in commodity agreements. On the whole, the United States has remained true to its free trade views and has refrained from giving in to these forces. Yet these concerns have shown themselves in management of the strategic stockpile in conflicts between foreign and domestic policy concerns.

When, for example, a large rise in tin prices stimulates a domestic demand to release tin from the stockpile to inhibit the price rise, the question arises, What will this do to U.S. friends in Malaysia? These conflicts in objectives have been a continual obstacle to economically efficient management of the stock.

Law of the Sea. A third area in which the episodic concern has risen with nonfuel minerals—in this case, particularly metallic minerals—has been in the drawn out Law of the Sea negotiations. Like the stockpile, this is a situation of conflict between domestic and foreign policy interests. The interests of American business have been in maintaining a regime at least broadly similar to one of private property rights in undersea mining, which would allow U.S. firms to exploit minerals to the extent that markets and technology permit. The developing nations would like to internationalize the rights in a "Global Authority," to which all member-nations would be required to make available their relevant technology. The authority would license undersea minerals development and use its income to assist economic development. Most nations have agreed to that proposal, in exchange for agreement on the extent of territorial and economic zones and rights of free passage in narrow waters. The Carter administration made the same choice, but the Reagan administration rejected it. The decision was made in part on the grounds that U.S. firms were the sole masters of the technologies that would permit exploitation of deepsea mineral deposits, that the proposed organizational structure led to unacceptable precedent-setting features, and that the country could get the other benefits of the new Law of the Sea regime without becoming party to the agreement. This stance is obviously unpopular with the developing countries, especially so in light of the long years spent in forging an agreement and in securing what was thought to be U.S. concurrence to it.

Food

The third global resource—food—has played a more continuous role in U.S. foreign policy than have the nonfuel minerals, but it, too, clearly is secondary. With few exceptions, domestic interests and concerns have dominated international considerations in shaping U.S. food policy. The major themes of policy have been restricting imports and promoting exports, both obviously reflecting the interests of domestic producers and their continuing political strength. As Dale Hathaway indicates in chapter five, however, U.S. agriculture increasingly is influenced by foreign markets and international events.

Exports and imports. U.S. farmers are highly efficient producers of the major food export crops—grains and soybeans. In the last decade and a half, U.S. exports of agricultural products have grown nearly seven times, and the net balance of U.S. agricultural trade more than twelve times. In a period of trade deficit, the contribution of agricultural exports to the trade balance has grown steadily, at least until 1981–82. Thus, the obvious U.S. interest in a continuing commercial diplomacy directed at removing further barriers to imports of U.S. food products, especially in Western Europe and Japan.

On the import side, the United States itself has maintained a variety of barriers, the most important of which have concerned dairy products, meat, and sugar. The barriers to imports of dairy products and meat are in the form of health regulations, and thus ostensibly not a matter of foreign policy. But many barriers imposed on imports from the United States by other countries also appear in forms other than tariffs and import quotas, and U.S. officials hardly are ready to accept them at face value as reflecting only health or safety concerns and the like.

The sugar situation is more straightforward. For a long time the United States has subsidized domestic beet sugar production heavily, and restricted by quota imports of the cheaper cane sugar from semitropical producers. The allocation of the import quota among producers has been used regularly as an instrument of foreign policy, to reward friends, such as the Dominican Republic, and punish enemies, such as Castro's Cuba.

Surplus management. For a long period after World War II the major instrument of domestic agricultural policy was price support of the chief cereal crops, cotton, and tobacco, maintained by government purchase of (or loans against) that part of the crop that could not be sold at or above the support price. The result was an accumulation of grain and, to a lesser degree, raw cotton and tobacco in government stocks. This in turn led to a policy of using surplus crops as a large part of U.S. economic aid to developing countries, first under Public Law 480, and then under the slogan, "Food for Peace."

The domestic purpose of these programs has been clear, and they have enjoyed broad political support. But their consequences for the recipients have been a matter of some controversy. It generally is agreed that untied aid equivalent to the world market value of the food would have been more helpful to the food-short countries. Many further believe that food aid has acted as a disincentive to local agriculture in recipient countries, reinforced governments in policies that were inimical to agriculture, and thus delayed progress toward food self-sufficiency that they otherwise could have achieved.

U.S.–Soviet relations. The exceptions to the broad dominance of domestic agricultural interests in shaping food policy have centered on U.S. relations with the Soviet Union. In 1966–67—an early marker of the beginning of détente—the United States negotiated its first grain sale with the Soviets. To recapitulate Hathaway's discussion, the full blooming of détente led to the signing, in 1975, of the first five-year U.S.–Soviet agreement on grain sales. That agreement, of course, fell victim to the Soviet invasion of Afghanistan in 1979 and President Carter's decision to restrict—but not eliminate—grain exports to the USSR as a sanction. Two years later, in April 1981, President Reagan lifted the Carter partial embargo, belatedly fulfilling a campaign promise, despite the continuing occupation of Afghanistan and the fact that U.S.–Soviet relations in general had worsened. Lifting the embargo, however, was as far as the Reagan administration then wished to go: the United States offered no new long-term agreement, and Soviet short-term grain purchases from the United States were significantly lower than they had been before the partial embargo.

In December 1981, the Polish government's declaration of martial law in its effort to suppress the Solidarity trade union led to the suspension of U.S. government credit for Polish grain purchases, and the further American refusal to negotiate a new long-term agreement for Soviet grain purchases.

Finally, after two one-year extensions of the lapsed agreement, the Reagan administration and the Soviet Union, now under Yuri Andropov, in August 1983 concluded a new five-year pact. Its terms on the surface appear to be major improvements—from the American standpoint—over the initial agreement: both the minimum purchase level and the level allowable without consultation are 50 percent larger than before. But those levels, now 9 and 12 million metric tons, respectively, still are small relative to the amounts the United States wishes to sell and to what the Soviets actually have purchased in recent years. Thus, the new agreement may be interpreted as an expression of Soviet caution: twice burned by embargoes, the Soviets probably do not believe they can rely fully on what must appear to them as a capricious—not to say arbitrary—United States. And they can afford to be wary, as Soviet agricultural production is picking up after four dismal years and especially as they have effectively diversified their sources of external supply, with Canada and Argentina playing prominent roles.

Thus, U.S. food policy confronts a self-imposed dilemma. American agriculture increasingly is oriented to the export market, but important parts of that market from time to time are declared off limits for foreign policy reasons. As Hathaway concludes, whatever shape food policy takes almost certainly will be influenced heavily by international events only partially—if at all—under the control of U.S. policymakers.

Through a Glass, Darkly

So much for what has happened. What can we expect will happen in the near future? Any specific prediction almost certainly will be wrong, but the best forecast is that the situation of the recent past will continue: the whole politicoeconomic system has fairly high inertia.

Oil Still Dominant

Everything in chapter two makes it plain that oil will continue to be a major security issue. It is possible that the relative importance of the Middle East as a source of European and Japanese imports will diminish as other sources are developed. But a counterbalancing possibility exists that Soviet production will decline sufficiently so that the Soviet Union, or rather the Soviet bloc, will become a net oil importer, adding to the pressure on world supply.

The current situation of oil surplus and falling real prices surely is transient. Even a sluggish recovery from the world recession of the early 1980s will boost demand sufficiently to make it likely that OPEC once again will be able to function effectively to limit the output of at least the Gulf producers to a level that sustains a constant or slowly rising real price.

Furthermore, the Middle East remains the source of "extra" supply—the producing area with the largest ratio of reserves to production and thus the most easily expansible output. Since social and political instability and the chances for international conflict in the area will continue to be strong, the potential for supply interruption and political use of the threat of interruption will remain high. Nor does this conclusion depend particularly on a pessimistic assessment of the prospects for resolution of the most sensitive and difficult issues between the Arab nations and Israel. Even a best-case scenario—Jordanian–Israeli negotiations, in which the Palestinians take part, that result in formal peace on three Israeli borders—does not warrant the expectation that the Middle East as a whole would achieve sufficient stability to defuse the oil threat. Inter-Arab rivalries, religious conflicts, and the potential for explosive internal political change within the area would remain.

Nonfuel Minerals: Focus on Southern Africa

Nonfuel minerals will sharpen U.S. foreign policy and security concerns as the situation in southern Africa grows more tense and interruptions of mineral supplies occur or appear near at hand. Specifically, tension will increase between the interests of U.S.-dominated and largely U.S.-owned corporations with mining operations in southern Africa and other U.S. foreign policy interests. Much of it probably will appear in the shape of a debate about where the "true" U.S. security interests lie, whether in maintaining current access to the mineral resources of the Republic of South Africa, or in good relations with the black African states on its borders. My own unconfident estimate is that this conflict will be resolved in favor of the latter, both for the expected reasons of domestic politics and, as Tilton and Landsberg suggest, because the particular mineral resources involved will not prove indispensable or unavailable elsewhere.

If we accept the prediction of most experts that undersea mining will not become economic before the end of the century, then conflicts over the new Law of the Sea and the U.S. relationship to the property regime in seabed minerals it establishes are unlikely to become significant foreign policy issues in the near future. Yet the possibility of renewed international contention over the issue—arising for ideological rather than practical reasons—cannot be ruled out.

Food: What's Past Is Prologue

Past patterns and relationships concerning food may be expected to persist, with producers' interests remaining the leading element in U.S. foreign economic policy in this area. A turn toward renewed détente in U.S. relations with the Soviet Union is not unlikely, though certainly not inevitable. Indeed, the conclusion of the long-term grain agreement once again may prove an early sign of a change in the overall tenor of U.S.–Soviet relations.

There probably will be repeated episodes such as those of 1973–74 in which the run up of international food prices has a disproportionate impact on poor importers, and the question of emergency relief will arise again. The possibility of the United States using food as a foreign policy weapon again may come to the fore, and the slogan, "a bushel for a barrel," even may be revived. Jingoistic chest-thumping aside, this will remain an unrealistic notion.

Although the United States currently accounts for a very large share of the world's exports in major grains, it simply does not control enough of the world's potential supply to make such a threat possible on its own. Nor do other major suppliers and potential suppliers share U.S. interests and views closely enough to make an effective "OGEC" likely.

Two Premises for the Future

What should the future be, if Americans could shape it as they wish? Or perhaps the question is better posed in terms of what can be done as a matter of U.S. national policy to help shape a better future.

Broadly and briefly, the answer in my view is to base U.S. policies on two premises: first, a continued commitment to an open world economy; and second, the recognition that energy, mineral, and food supplies are truly global resources. These premises lead directly to the conclusion that increases in global supply—wherever they occur—are in the long-run U.S. national interest, even though they may not be in the immediate interests of all or even particular U.S.-domiciled or U.S.-owned producers.

Some specific applications of the principle are clear. For example, it was inappropriate—not just politically unwise and futile—to try to hinder the Soviets in building the natural gas pipeline from Siberia to the western border of the Bloc. Will it not increase the supply of energy to Western Europe? Similarly, U.S. administrations of both political parties have blown hot and cold on the question of making available—or denying—to the Soviet Union and other unfriendly countries U.S. oil-drilling and -exploration technologies. The U.S. Export Administration Act allows the president to restrict exports for reasons of national security and foreign policy. As this volume goes to press, the act is being debated in Congress preparatory to renewal. Whatever form renewal takes, one hopes that President Reagan and his successors use it sparingly and not as an all-purpose foreign policy bludgeon.

Another recent example in the energy field was a proposed World Bank program to encourage oil exploration in the Third World. U.S. opposition killed the proposal, again in contravention of the principle of treating resources globally.

The issues of seabed mining in relation to the Law of the Sea Treaty are perhaps less clear-cut. It could be argued that the Global Authority will behave as a restrictive monopoly, limiting supplies to raise prices, and that a more open and competitive regime would be more in accord with the principle of maximizing long-run supply. On the other hand, an agreed peaceful regime—even a monopolistic one—may be more conducive to production than a situation of open conflict in which property rights rest on no agreed basis; the treaty regime may be the attainable second best. The United States may come to rue its rejection of the treaty.

Two food policies could well serve long-run U.S. interests, although they might diverge from the shorter-run interests of American producers. The first is an increase in both the scale and effectiveness of international programs to increase food production in the Third World, especially the capacity of poorer countries to raise the outputs of their own farms. This would require both more U.S. resources and leadership than have been given such efforts in recent years. Second would be coordination of efforts to create and maintain an international food reserve to deal with the extra burdens that sharp short-run fluctuations in production and prices place on poor importing countries.

Finally, the most important immediate obstacle to maintaining a truly global view is the recent worldwide recession and the low growth rates and high unemployment in the industrial economies that threaten to continue through the next cyclical recovery. These circumstances stimulate protectionism, make it politically difficult for the United States and other nations to take a global view, and make it much more likely that recent behavior will serve as the model for future behavior. U.S. interests in natural resources *are* global interests, but it is difficult to see this essential truth when suffering from national myopia.

Notes

1. The kind of thinking represented in the deliberations of the Paley Commission and in its final report, *Resources for Freedom,* was institutionalized in part through the establishment in 1952 of the sponsor and publisher of this volume, Resources for the Future. For a brief history of William S. Paley's role in creating RFF, see *Resources for the Future: The First Twenty-Five Years* (Washington, D.C., Resources for the Future, 1977).

2. A comprehensive history of the 1959–73 U.S. experience under the Mandatory Oil Import Program is provided by Douglas R. Bohi and Milton Russell in *Limiting Oil Imports: An Economic History and Analysis* (Baltimore, Md., The Johns Hopkins University Press for Resources for the Future, 1978).

epilogue

The Inevitability of the Unexpected

McGeorge Bundy

The preceding chapters have been thoughtful and on the whole persuasive accounts of the ways relationships among global natural resources and political behavior differ from what we tend to assume in most daily debates. We have learned also that each resource has its own characteristics not necessarily shared by others nor adequately described by such blanket terms as *international trade*. Can any common threads be discerned among what often appears as a tangled skein?

Economics Versus Politics

The first lesson I glean from my colleagues' offerings is that economics is usually more important than politics.

Milton Russell's searching account of the oil shocks of the 1970s demonstrated to me that very large underlying economic forces were at work in the making of the first of those shocks. Moreover, the way the shock was absorbed also is explicable primarily in economic terms, as the product mainly of market forces. Market forces did in fact cushion the blow—though at great overall cost—in ways that were less damaging than many of us had feared at the beginning.

I am struck particularly by the force of Russell's point that the 1974 price change well may be recorded in economic history, when it comes to be written, as an adjustment to the real value of property once the price setter had effective and reasonably durable political control over what was in the ground. That rings true to me and I find it instructive. It presses home upon us the fact that while a judgment on other nations' behavior may be by definition a political judgment, it had better be based, insofar as we can manage it, on an accurate appreciation of the relevant economic realities.

The same lesson emerges from the analysis of policies for nonfuel minerals and food. For example, two propositions about President Carter's post-Afghanistan grain embargo are obvious in retrospect. First, the embargo could not long survive in the face of the felt economic interests of a very large and powerful sector of U.S. society. Second—and this was apparently not clearly perceived in the White House at the time—it would not work in terms of the market behavior of major suppliers beyond the effective political influence of Washington. Thus, both straight economic forces and economic interests that express themselves through political pressure can have very large roles, roles that often overmatch directly political considerations such as those behind the grain embargo.

Yet we also have to recognize that economics is more important than politics, *except when it is not*. Some things are so heavily invested by political interest that it is only by rearranging political attitudes, and indeed national policy itself, that the economic interest will be served. Excepting agriculture, that is, I think, the history of U.S. trade relations with the Soviet Union, a relationship in which the economic interest has always been at the mercy of changes in the political atmosphere between the two countries.

Politics and economics interact in other ways, and it is sometimes hard to tell which matters more. But I would agree with an idea suggested by John Tilton's and Hans Landsberg's conclusion on strategic stockpiles: that in any effort where the underlying objective is to strengthen *national* economic policy, one should try to design the organization and management so that it is insulated from immediate interest-group pressures. We know from the history of the Federal

Reserve Board—to take a highly conspicuous case—how hard that is and also how ambivalent feelings can be if success is too great. Yet the need for insulation is real.

Arranging for Adaptability

The second general lesson I draw from the volume's analysis of economics as a determinant of policy is that while shocks as a general phenomenon are predictable, particular shocks and their timing—almost by definition—are not. That is the common lesson of the history of unpleasant surprises in oil, of pleasant surprises in food, and even of the absence of surprises in nonfuel minerals. We must expect the unexpected, and the problem thus presented is whether and how society can prearrange adaptability to the unexpected.

I would argue that my first proposition—on the importance of economics—prevails here, and that on the whole the market is the most powerful single instrument for allowing for adaptability. But my coauthors throughout the preceding pages have documented a kind of yes–but—that the market pays imperfect attention to the present value of reducing future risk, and also that it resists playing its full part in the compensation for some unexpected shortfall. Thus, we have seen roles proposed for nonmarket forces—that is, governmental intervention—of several kinds. Two are worth special emphasis.

Stockpiles. The first is stockpiles. Both for oil and for strategic minerals it is hard to beat the stockpile on cost and availability as a means of providing, in advance, a cushion against future shocks.

In a strong speech delivered in October 1982 in South Africa, former World Bank president Robert S. McNamara laid out sharply his own view that a time of crisis looms in the less than ten-year future between the United States and a South Africa that does not meanwhile change its racial policies dramatically. McNamara called for the immediate and large-scale establishment of the necessary stockpiles of the minerals that would be affected if there were any attempt to influence our policy by embargo from South Africa. Tilton and Landsberg see a similar threat through the inability of a South Africa embroiled in domestic strife to produce or export minerals. These are only examples of this kind of danger, although my guess is that among nonfuel minerals those of southern Africa, taken as a whole, are the ones most likely to require this kind of protection in advance.

Ironically, the broad agreement on the value of the strategic oil reserve is matched by a general recognition of the low priority that enterprise is accorded in the pull and haul of annual budget making. Even as an outside amateur observer I have seen a half-dozen cases since 1973 in which one of the easy targets at the final pinch is a stockpile whose time of need simply cannot be predicted.

It is of course impossible to know with any degree of certainty what might have happened in 1979 if, at the time of the interruption of Iranian supply, there had been a billion-barrel U.S. strategic petroleum reserve. I do not know that the result necessarily would have been good, but it does appear safe to predict that it would not have been worse. And it is at least possible that an effective reserve would have permitted some diplomatic flexibility and a relatively coherent response from the industrial countries rather than collective panic and another doubling of prices.

I am not an economist, but I have the impression from Russell and others that there may be a considerable difference between the oil price shock of 1974 and that of 1979–80 in terms of their relative real economic justification. If that is right, then the value of a reserve might have been correspondingly greater in the second case, since crisis expectations may be easier to moderate than underlying economic pressures. But all I really am arguing is that if a third shock is possible or likely at some point in the 1980s, it would be pleasant to have more opportunities than we do now, with the strategic reserve falling victim once again to immediate budget pressures.

Research and development. The second general role that appears to be appropriate for the U.S. government is support for research and development. In the golden age of the years immediately after World War II, research and development was a splendid flag behind which all could rally. Vannevar Bush could write about "the endless frontier,"[1] and the nation could launch a National Science Foundation and other instruments of scientific support on the wave of popularity generated by the undoubted achievements of wartime science. Research and development is a somewhat more battleworn flag now. We have learned more about the difficulties of doing it right, doing it without waste, figuring out which parts of it belong to the public sector and parts of it should be assigned to or imposed upon those who hope for the rewards of their work.

Nonetheless, the importance of a governmental role in this area is clear to me and is emphasized explicitly or implicitly by the other authors. Accordingly, it is puzzling that those in power often do not see the merits of this kind of governmental investment in the nation's future. Indeed, the connection between government and basic research has to be remade over and over again. It is sad but by no means surprising that it should have to be made once more. Some of the most poignant passages in George Kistiakowsky's extraordinary memoirs of his time in the Eisenhower White House have to do with the way in which he and his predecessor, Jim Killian, persuaded a president who was no automatic believer in strong government or in large budgets that support for basic research and related parts of higher education were in fact major and necessary activities for the national government.[2]

Surely one of the truths of this volume is that this role still is paramount. Recognizing all the hazards and complexities it entails, we must encourage federal participation in the kind of research and development that enlarges our choices as we face not only the predictable, but also the unpredictable changes of the future.

A Calculus of Political Economy

A third lesson: granted the high role of economics and granted the necessity, at certain edges, of political decision, there is a requirement for what I will call, for want of a better word, a calculus of political economy, and one which goes further than economics as such or political interest as such.

For example, one way to understand the value of a stockpile is to compare it to the cost of a protective military deployment to a threatened area. The same kind of calculation, which is clearly not one for which there is a unit of account readily available, needs to be applied to, say, the relative value of the economic impact on the Soviet Union of a U.S. denial of pipeline technology, and the impact of that same enterprise on the political coherence of the Atlantic alliance. If government avoids that calculation, or neglects it, it can shoot itself in the national foot for the sake of noble but uncertainly effective principles.

In this wider process, there is a need for kinds of communication at which the U.S. government is not very good—communication not only with itself, but also with those who have to make decisions and take choices in the private sector.

Government must be sensitive to and understanding of, though not necessarily in agreement with, all kinds of groups whose attitudes have a bearing on decisions that in themselves bear back on the prosperity, the interests—the legitimate interests—of many sectors of American society. Individual interests are fairly adept at representing themselves at the Cabinet level, as in commerce, labor, and agriculture, but governmental structure and tradition militate against bringing such considerations together where a decision must be made that takes account of many kinds of interests. And the failing here is not mainly on the side of the interest groups.

The Soviet Factor

In this calculus of political economy, in the adjustment of political to economic interests, in the recognition that unpredictable shocks are predictable but open to moderating action, and in thinking about the diversity of the interests that are affected and of the instruments that are available to the United States, we should bear in mind that all of this affects and is affected by our picture of the USSR, and its picture of us. It is no mere happenstance that each of the preceding chapters emphasized the importance of U.S.–Soviet relations.

Popular in some powerful circles of American thinking is what I will call the monolithic picture of the Soviet Union, a picture in which it is assumed that this vast malign force is directing its undivided attention to the discovery of U.S. weaknesses, the penetration of U.S. gaps, the exploitation of U.S. failings, the avoidance of U.S. strengths, and the destruction of all U.S. hopes. I suggest this view is off the mark, because it assumes that the Soviets think about all issues only in terms of what will hurt the United States.

If I might confine the analysis simply to the thermonuclear issues, this view makes the very mistake about the Soviet Union that it seemed to make—but did not make—about the United States back at the very beginning of the Cold War. Recall that the United States had a monopoly on the bomb, and that the Soviet Union accused America of trying to exploit that monopoly by "atomic diplomacy." But at the same time, by its own real behavior, its own firm refusal to admit that atomic weapons conferred a decisive diplomatic power on the United States, the Soviet government showed that it did not believe its own propaganda. I know of no sober student who believes today that atomic diplomacy served any serious American interest or indeed was ever seriously tried, although it certainly was briefly a temptation to Secretary of State James Byrnes. It was not in fact a plausible instrument of pressure on the Soviet Union, but at best one element in the deterrence of open aggression.

I think that the underlying Soviet estimate then was correct, that the U.S. failure to gain major diplomatic leverage from its atomic monopoly was essentially natural. The U.S. monopoly had an undeniable effect on the military balance, but there was a fully countervailing balance in the other kinds of strength available to the Soviet Union. I think it is even more evident today that at the levels that strategic weapons now have reached, approaching 10,000 on each side (somewhat more for us than for them, but somewhat bigger for them than for us), thousands of which would survive any strategic first strike by either side, the notion that either side has an exploitable strategic nuclear advantage is, to put it bluntly, nonsense.

I believe the government of the Soviet Union is more intelligent on this point than are those Americans who assert a significant Soviet superiority. It is quite simply a wrong picture of the relationship to suppose that the United States now suffers from strategic inferiority, and still less that it is mortally endangered by some hypothetical threat to the Minuteman missiles that represent less than a quarter of the U.S. strategic force.

It appears to me that, instead of engaging in the fancy that the Soviet Union is moved by no other force than total hostility to the United States, we should think about it as I dare to venture that it thinks about itself: as a large, strong, durable force, equal in some large sense to the United States (an equality that matters enormously to it); beleaguered by persistent economic failure of a sort that would be deeply disruptive to a less powerfully totalitarian regime; surrounded by nonfriends in large numbers; embroiled in an adventure in Afghanistan that has not come out as intended and that was not a part of some remorseless plan to march toward the Mediterranean; and bedeviled by the uneasiness of conquest in Poland.

The world is diffuse, and the instruments of both contest and cooperation are complex. The advantages to the United States of trading with, talking with, and dealing with the Soviet Union are advantages *to Americans*. They exist in trade as they exist in dealing with the nuclear danger.

The Inevitability of the Unexpected

A final summary point and, like the last, one that is directed not just to economists and political scientists and the "policy community" but to everyone—to government, business, interest groups, and citizens at large. That is, all of us Americans are responsible for recognizing the inevitability of the unexpected, and the consequent requirement to be adaptable. This recognition is found more often on the sidewalks of New York or Seattle or Albuquerque than in the corridors of bureaucracy, but as a people we are in fact highly adaptable. To one degree or another, adaptability permeates the working level all through our society, though it is constantly in conflict with its opposite—a refusal to anticipate or to respond to even the more predictable forms of change.

The ability to adapt, in all sectors of our society, is measured at least in part by our readiness to expect the unexpected, our capacity not to be taken by surprise. I believe that everything in this book—indeed, everything we know more generally about the hectic rate of change in modern life—should lead us to put more emphasis than ever on strengthening the kind of respect for the certainty of the unexpected that can reinforce our individual and national readiness to adapt to whatever the future has to offer.

Notes

1. Vannevar Bush, *Science—The Endless Frontier: A Report to the President on a Program for Postwar Scientific Research* (Washington, D.C., National Science Foundation, 1945; reprinted May 1980).

2. George B. Kistiakowsky, *A Scientist at the White House: The Private Diary of President Eisenhower's Special Assistant for Science and Technology* (Cambridge, Mass., Harvard University Press, 1976).

About the Contributors

McGeorge Bundy, professor of history at New York University since 1979, for the previous thirteen years was president of the Ford Foundation. He was associated with Harvard University for a dozen years, most of them as dean of the Faculty of Arts and Sciences, and in the early 1960s served as special assistant for national security to Presidents John F. Kennedy and Lyndon B. Johnson. In 1982, he was a co-winner of the Albert Einstein Peace Prize.

Emery N. Castle is president and senior fellow at Resources for the Future. He came to RFF in 1976 from Oregon State University, where he had been professor and head of the Department of Agricultural Economics, dean of faculty, and dean of the Graduate School. He has been president of the American Agricultural Economics Association and chairman of Oregon's Water Policy Review Board, among other public and professional duties. He is co-editor of *U.S.–Japanese Agricultural Trade Relations,* published by RFF in 1982.

Dale E. Hathaway is principal and vice president of the Consultants International Group, Washington, D.C. From 1977 to 1981 he was assistant secretary, then under secretary for international affairs and commodity programs at the U.S. Department of Agriculture. He has been director and research administrator of the International Food Policy Research Institute, program advisor to the International Division of the Ford Foundation, and professor and chairman of the Department of Agricultural Economics at Michigan State University. He has served as president of the American Agricultural Economics Association.

Carl Kaysen is David W. Skinner Professor of Political Economy and director of the Program in Science, Technology, and Society at the Massachusetts Institute of Technology. From 1966 to 1976 he was director of the Institute of Advanced Study at Princeton, New Jersey, and for the previous sixteen years he was a member of the Economics Department at Harvard University, where he also served as associate dean of the Graduate School of Public Administration. His extensive public service includes two years during the mid-1970s as chairman of the National Academy of Sciences' Committee on Scholarly Exchange with the Soviet Union.

Hans H. Landsberg is a senior fellow and former director of the Center for Energy Policy Research at Resources for the Future. He joined RFF in 1960, was coauthor of one of its first major publications, *Resources in America's Future* (1963), and directed its programs of resource appraisals and of energy and materials. A consultant to many governmental and nongovernmental organizations, he was chairman of the Ford Foundation-sponsored study group that produced *Energy: The Next Twenty Years* (1979). He is coauthor of *Energy Today and Tomorrow: Living with Uncertainty* (1983). In 1980 the American Institute of Mining Engineers awarded him the annual Mineral Economics Award, and in 1983 he received the International Association of Energy Economists' award for contributions to the profession and its literature.

Kent A. Price is associate director of public affairs and assistant to the president at Resources for the Future. He is the principal writer and editor of RFF's *Annual Report* and other institutional materials, co-editor of its quarterly, *Resources*, and editor of *Regional Conflict and National Policy,* published by RFF in 1982. Before joining RFF in 1975, he was chief writer for the president of the University of California.

Milton Russell is special assistant to the administrator at the U.S. Environmental Protection Agency. He was associated with Resources for the Future from 1976 to 1983 as a senior fellow and served as director of RFF's Center for Energy Policy Research for the last four years of his tenure there. He was a senior staff economist with the President's Council of Economic Advisors during 1974–76 and, among several academic posts, was professor and chairman of the Department of Economics at Southern Illinois University. He is coauthor of, most recently, *Limiting Oil Imports: An Economic History and Analysis* (1978), and *Energy in America's Future: The Choices Before Us* (1979), both published for Resources for the Future.

John E. Tilton is professor of mineral economics at The Pennsylvania State University and co-director with Hans Landsberg of the Mineral Economics and Policy Program at Resources for the Future. He is the coauthor of *Public Policy and the Diffusion of Technology* (1978), author of *The Future of Nonfuel Minerals* (1978), and editor of *Material Substitution: Lessons from Tin-Using Industries,* published by RFF in 1983. Currently on leave from Penn State, he is director of the Minerals Trade and Markets Project at the International Institute for Applied Systems Analysis, Laxenburg, Austria.

Mary Beth Zimmerman, a research assistant at Resources for the Future, has completed graduate work in economics at The American University, Washington, D.C. She is coauthor of "An Update of Econometric Studies of Energy Demand," and is a contributor to *Natural Gas Markets After Deregulation,* published by RFF in 1983. Her other research at RFF includes a study of industrial electricity demand in the Tennessee Valley Authority area and analysis of petroleum exploration and development in developing nations.

Index

For Product Safety Concerns and Information please contact our
EU representative GPSR@taylorandfrancis.com Taylor & Francis
Verlag GmbH, Kaufingerstraße 24, 80331 München, Germany